Springer Tracts in Modern Physics 97

W0112334

Editor: G. Höhler
Associate Editor: E. A. Niekisch

Editorial Board: S. Flügge H. Haken J. Hamilton
H. Lehmann W. Paul

Springer Tracts in Modern Physics

* denotes a volume which contains a Classified Index starting from Volume 36.

Wolfgang Ehrfeld

Elements of Flow and Diffusion Processes in Separation Nozzles

With 73 Figures

Springer-Verlag Berlin Heidelberg GmbH 1983

Dr. Wolfgang Ehrfeld

Institut für Kernverfahrenstechnik, Kernforschungszentrum Karlsruhe, Postfach 36 40
D-7500 Karlsruhe, Fed. Rep. of Germany

Manuscripts for publication should be addressed to:

Gerhard Höhler

Institut für Theoretische Kernphysik der Universität Karlsruhe
Postfach 6380, D-7500 Karlsruhe 1, Fed. Rep. of Germany

*Proofs and all correspondence concerning papers in the process of publication
should be addressed to:*

Ernst A. Niekisch

Haubourdinstrasse 6, D-5170 Jülich 1, Fed. Rep. of Germany

ISBN 978-3-662-15744-2   ISBN 978-3-540-39513-3 (eBook)
DOI 10.1007/978-3-540-39513-3

Library of Congress Cataloging in Publication Data. Ehrfeld, Wolfgang 1938 –. Elements of flow and diffusion processes in separation nozzles. (Springer tracts in modern physics; 97) Bibliography: p. Includes index. 1. Isotope separation. 2. Nozzles — Fluid
dynamics. 3. Diffusion. I. Title. II. Series. QC1.S797 vol. 97 [TK9360] 539s [621.48'37] 82-17008

© by Springer-Verlag Berlin Heidelberg 1983

Originally published by Springer-Verlag Berlin Heidelberg New York in 1983.

Softcover reprint of the hardcover 1st edition 1983

2153/3130 – 543210

Preface

Around the mid-fifties, E.W. Becker observed a significant mass separation in a molecular beam apparatus which he had constructed to study the effects of quantum statistics upon gas kinetic transport phenomena. This accidental observation immediately initiated a series of careful experiments, and it was found that the mass separation resulted from the expansion of the gas in the aerodynamic stage of the molecular beam apparatus. From the very beginning, the efforts of Becker and his co-workers were directed at finding a technical application of the so-called separation nozzle and, consequently, at separating uranium isotopes. More than a quarter of a century after that first observation, this goal has been achieved: within the framework of the German-Brazilian agreement on cooperation in the peaceful use of nuclear energy, a separation nozzle demonstration plant for industrial-scale production of light water reactor fuel will be erected in Brazil.

The present monograph is an attempt to systematize the results of studies performed by the author and his colleagues on the physics of the separation nozzle process, which is based on pressure diffusion in curved flows of uranium hexafluoride and a light auxiliary gas. The description of the complex flow and diffusion in separation nozzles includes transient separation effects in ternary mixtures of gases, as well as non-equilibrium phenomena typical of low-density flows of disparate mass mixtures. In addition, the principles and designs of the most important types of separation nozzles are explained in detail. In view of the practical application of the separation nozzle process, general correlations are pointed out between the physics of aerodynamic separation and the technical expenditure for large-scale production of enriched uranium.

It is a pleasure for me to acknowledge my indebtedness to Professor E.W. Becker, my teacher and constant source of helpful and stimulating guidance. I am very grateful to my colleagues at the Institut für Kernverfahrenstechnik, whose experimental and theoretical investigations provided the basis of this monograph; in this connection, I am indebted to Dr. P. Bley, Dr. W. Berkhahn, Dr. H. Brandstädter, Dr. H. Breton, Dr. R. Dürr, Dr. G. Eisenbeiß, Dr. U. Heiden, Dr. U. Knapp, Dr. G. Krieg, Dipl.-Phys. W. Schelb, Dr. E. Schmid and, in particular, to my wife, Dr. Usula Ehrfeld.

December 1982   *Wolfgang Ehrfeld*

Contents

1. Introduction

The separation of isotopes is a process of fundamental importance in many areas of science and technology. Since the basic effects capable of being used for this purpose have different efficiencies, depending on the type of isotope or the quantity required, a multitude of isotope separation techniques exist today, which meet the broadest applications. Normally, a distinction is made between the so-called statistical methods, in which the difference in the mean values of a physical quantity is used for the separation process, and the non-statistical methods, in which the individual differences in nuclear masses or nuclear volumes of isotopes directly determine separation /1/.

The most widely known statistical methods are based on gas kinetic transport processes, such as pressure diffusion, thermal diffusion or molecular effusion (diffusion methods) and on differences in the equilibrium distribution of isotopes among various phases (phase equilibrium methods) or chemical compounds (chemical exchange methods). Among the non-statistical methods which, unlike the statistical methods, in principle would allow the pure material to be produced in one step, there is particularly the separation of ionized isotopes and isotopic compounds in magnetic or electric fields (mass spectrometer); there is also the possibility to make use of differences in the atomic or molecular spectra for excitation of specific isotopes and separation of the type of isotope selectively excited by subsequent ionization, dissociation or chemical conversion (photoselective methods).

The separation of isotopes is carried out on an industrial scale for the production of heavy water and, in particular, for the enrichment of the light uranium isotope ^{235}U /1,2/. This isotopic species which is fissionable by thermal neutrons, occurs in natural uranium only with an abundance of 0.72 mole %. Since modern nuclear power plants are mostly equipped with light water reactors which require uranium fuel with ^{235}U contents of 2% - 4%, uranium isotope separation is obviously an essential step in the peaceful use of nuclear energy.

The requirements of enriched uranium at present are still covered mainly by facilities using the gaseous diffusion method /3/. This process is based on the mass dependence of molecular effusion through porous membranes and is carried out with gaseous uranium hexafluoride. However, it can work economically only in extremely large plants; this is ultimately due to the small change of the isotopic composition in the elementary separation process at the membrane and the correspondingly large flow of material, which results from the multiple repetition of the elementary separation process in a cascade. The most important competitor of the gaseous diffusion process at present is the gas centrifuge process /4/, in which gaseous uranium hexafluoride is fed into a cylinder rotating at high speed and use is made of pressure diffusion for isotope separation; the process is being applied successfully on a technical scale in a number of pilot plants and demonstration facilities. Other methods of uranium enrichment developed with great financial and technical expenditures are the separation nozzle method /5/, the UCOR process based on the vortex tube /6/, chemical exchange processes /7/, and photoselective separation by selective excitation of isotopes in uranium vapor or gaseous uranium compounds by means of laser radiation /8/.

The most important alternative method relative to the gaseous diffusion and centrifuge processes today is the separation nozzle process developed at the Karlsruhe Nuclear Research Center /9-12/. It is characterized by simple technology and the possibility of economic operation even in relatively small plants. In the separation nozzle process, the pressure gradients and inertia forces in a curved gas flow containing uranium as UF_6 are utilized for separating the uranium isotopes. Such separation principle was first proposed in the Second World War by DIRAC for enriching ^{235}U. However, the first tentative experiments performed on model gas mixtures at that time were not very promising, and the premature conclusion was drawn that aerodynamic separation methods were unsuitable for uranium enrichment on a technical scale /13/. Around the mid-fifties, BECKER began his first studies on separating uranium isotopes in UF_6 gas jets freely expanding from a convergent nozzle and split by a skimmer or knife edge into a core enriched in the heavy isotope and an outer shell enriched in the light isotope /14/. The elementary effect of isotope separation observed in this case was only very little above that of the gaseous diffusion method, even if the UF_6 was strongly expanded, which made technical application of such arrangements hardly promising /15/.

The breakthrough in the development of the separation nozzle was achieved by BECKER in the early sixties when, instead of pure UF_6, a mixture of UF_6 and a light auxiliary gas in a high molar excess was used /16/ and free expansion was replaced by a guided deflection of the flow along a curved wall /17/. These measures helped to attain a much higher flow velocity and greater deflection of the flow than in the free expansion of pure UF_6. The elementary effect of uranium isotope separation was raised considerably above that of the gaseous diffusion process, while simultaneously reducing the expansion ratio required for economic operation of the nozzle. As a consequence of the improvement in performance data, the technical expenditure for industrial uranium enrichment by the separation nozzle method reached economically attractive levels /18-22/.

Since 1970, STEAG of Essen have participated at their own expense in industrializing the separation nozzle process /23/. In 1975, it was agreed within the framework of the German-Brazilian Nuclear Energy Agreement to carry out jointly the further development and commercialization of the process with the participation of industries and research establishments in the two countries /24/. At present, a first cascade section of a demonstration plant is being built. Its planned capacity of 300 000 SWU/a (SWU/a = kg separative work unit per year) is to ensure the supply of nuclear fuel to light water reactors with a total power output of roughly 3000 MWe /25/.

Along with the technical implementation of the separation nozzle method, which is carried out jointly with industry /26/, and testing of process components on a technical scale /27,28/, broad fundamental studies are being performed at the Karlsruhe Nuclear Research Center on the physics of the separation nozzle method.[1] These studies serve to elaborate approaches towards further improvements in the performance of the process and prepare their practical implementation by laboratory scale experiments. The success achieved so far in this physics development work is demonstrated by the fact that the power consumption of the separation nozzle method for uranium enrichment has been cut to half the level of 1970 /20,23/. The specific power consumption of an industrial separation nozzle plant is 2800 kWh/SWU, which means that about 3% of the electricity generated by the enriched uranium in light water reactors is to be expended on the enrichment process. The number of enrichment steps required to produce nuclear fuel has also been reduced by some 50%, so that at present a cascade arrangement of about 200 separation stages is sufficient to convert natural uranium into enriched material with 3 mole % ^{235}U. Also the other technical expenditures involved in uranium enrichment by the separation nozzle have been reduced considerably as a result of improved physical performance. There is no doubt that the separation nozzle process is developing into a real alternative to the established methods for the enrichment of the light uranium isotope /25/.

This monograph presents a survey of the basic physics underlying the separation nozzle method. After a short introductory description of the separation nozzle principle, some general correlations will be established between the operating and performance data of the separation nozzle on the one hand, and the technical expendi-

---

[1] Cf. survey articles /2,5,9,29-34/; studies of specific physical problems are cited below at the appropriate points in this monograph.

ture required for uranium enrichment on the other hand. This is followed by a detailed analysis of the separation and flow processes in the separation nozzle, which will deal especially with the effect of the light auxiliary gas on uranium isotope separation and with the flow phenomena characteristic of the separation nozzle in the transition regime between continuum and free molecular flows. This analysis, in turn, is used as a basis for explaining, in physical terms, the influences of various operating parameters of the separation nozzle on uranium isotope separation and on the specific technical expenditure. Since the complicated flow and separation processes in the separation nozzle can only partly be assessed theoretically and, for this reason, experimental methods dominate in the elucidation of physics events, the measuring techniques developed for these studies will also be described. In the final section of the study, special physics phenomena and some technological aspects of the most important embodiments of the separation nozzle will be treated.

2. Basic Principles of the Separation Nozzle Method

As with most separation methods applied on a technical scale, the separation nozzle method is a continuous process in which the mixture to be separated is fed as a steady-state flow to a separating element and split into fractions of different compositions. The components of the mixture are separated in the gaseous phase, the mixture being accelerated by expansion in a nozzle and deflected in the process. The centrifugal forces occurring during deflection and the resulting pressure gradients in the flow field cause a partial separation of the components of the mixture, as in a centrifuge. The heavy components are concentrated at the periphery, the lighter components in the inner regions of the centrifugal field generated by the curved flow. Depending on the type of separation nozzle, the curvature of the streamlines is brought about by

- deflection of a gas jet at a solid wall,
- mutual deflection of several gas jets,
- free expansion,
- partial stagnation of a flow,

or a combination of these measures (Sect.8).

In uranium isotope separation, the process gas used is a mixture of uranium hexafluoride (UF_6) and an auxiliary gas of low molecular weight (helium or hydrogen). The fraction of the light auxiliary gas in the process gas mixture is around 95 to 98 mole %, thus resulting in values between 9 and 21 for the mean molecular weight of the mixture. Because of the low mean molecular weight, much higher exhaust velocities are attained than in the expansion of pure UF_6 and, consequently, higher separating centrifugal forces are reached in the curved flow.

In addition to the acceleration of UF_6, another effect of the auxiliary gas has a positive impact on separation of the uranium isotopes /35/. This is based on the fact that the UF_6 molecules containing the heavy isotope will migrate more quickly to the periphery of the centrifugal field than the UF_6 molecules containing the light isotope because of the higher pressure diffusion velocity of the heavier species. As a consequence, an isotopic distribution will be passed before reaching equilibrium condition, at which the heavy isotope is already concentrated at the periphery of the centrifugal field, while the light isotope is still spread over a broader range. This transient state is characterized by a higher degree of isotope separation than the equilibrium condition, in which a steady-state Boltzmann distribution has been established for each type of molecule.[2]

The spatial change in the isotopic ratio in the field of the separation nozzle is determined by the ratio of the diffusive transport of the isotopes perpendicular to the flow direction to the total transport of isotopes in the flow direction. This ratio of transport streams has a finite value only in the transition regime between continuum flow and free molecular flow, since the necessary preconditions of low throughput and high-pressure diffusion stream can be met simultaneously only in this regime. The optimum Knudsen number of the separation nozzle flow for isotope separation purposes, i.e., the ratio of the mean free path of the molecules to a characteristic dimension of the nozzle, is 10^{-2} /33/.

Figure 2.1 is a cross section of a slit-type separation nozzle system, which has been the basis of most technical development work so far performed on the separation nozzle process. It will be referred to below as "standard separation nozzle". A gas jet containing 2 to 5 mole % of UF_6 and 98 to 95 mole % of H_2 or He is deflected at a fixed curved wall and expanded to roughly half the value of the nozzle inlet pressure, the heavy components of the mixture being concentrated at the deflection wall and the light ones in the inner regions of the flow. At the end of the deflection, the

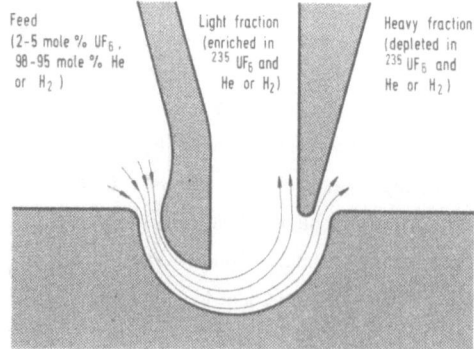

Feed
(2-5 mole % UF_6,
98-95 mole % He
or H_2)

Light fraction
(enriched in
$^{235}UF_6$ and
He or H_2)

Heavy fraction
(depleted in
$^{235}UF_6$ and
He or H_2)

<u>Fig.2.1.</u> Cross section of a slit-type separation nozzle system (standard separation nozzle)

<hr>

[2] A similar mechanism is applied in some types of flow classifiers and separators, in which the different settling rates of particles of a mixture of solids in a liquid or a gas are used for separation. Also in this case, a higher degree of separation is achieved during sedimentation than at the end of the sedimentation process /36/.

partly separated gaseous mixture is split by a skimmer into a heavy fraction depleted in $^{235}UF_6$ and the auxiliary gas, and a light fraction enriched in $^{235}UF_6$ and the
auxiliary gas. Under the operating conditions prevailing in practical applications
of the separation nozzle method, the relative difference in isotopic ratios between
the light and the heavy fractions is about 1.5%. For this reason, the elementary separation process must be repeated several hundred times in a cascade arrangement of
separation nozzle stages in order to raise the molar fraction of the light isotope
from 0.0072 in natural uranium by a factor of about 4 to the level of 0.03 required
for light water reactors (cf., e.g., /2,5,20,25,27/).

3. Characteristic Parameters of the Separation Element and Specific Expenditure

In a physical analysis of the separation process and for economic assessment of the performance of separation nozzle systems, certain parameters are used in analogy with those employed in other uranium enrichment techniques. These parameters, in a general way, take into account the fact that a given separation problem must in principle be solved by multiple repetitions of the elementary separation process in a cascade /37-40/. They describe the concentration change achieved in a single separation element, the splitting of mass streams and the expenditure specific to the process, and are explained below on the basis of the flow diagram shown in Fig.3.1 for a separation nozzle stage, which is the basic technical processing unit of a separation nozzle cascade.

The gas stream L fed to the separation nozzle system is composed of a partial stream flowing up the cascade and a partial stream flowing down the cascade, each having the same isotopic abundance. L is compressed to the feed pressure p_0 by means of a compressor and, after removal of the compression heat in a gas cooler, split in the separation element into a light fraction θL, which is enriched in $^{235}UF_6$ and auxiliary gas, and a heavy fraction $(1-\theta)L$, which is enriched in $^{238}UF_6$ and UF_6. The shift in concentrations between the light and the heavy fractions is characterized, as usual, by the separation factor A or by the elementary effect ε_A of the separation process. In most cases, it is appropriate for physical analysis of the separation process to express these quantities by means of the partial cuts θ_i. The

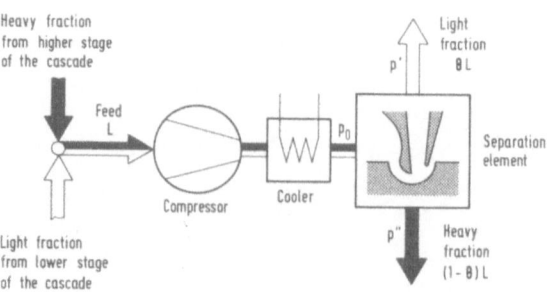

Fig.3.1. Basic flowsheet of a separation nozzle stage

partial cut of a component i of the mixture is the percentage of its throughput through the separation element, which is withdrawn in the light fraction stream. If θ_l and θ_h are the partial cuts of the UF$_6$ molecules containing the light and the heavy uranium isotopes, respectively, the following relationship results for the isotope separation factor A_{is} and the elementary effect ε_A of isotope separation:

$$\varepsilon_A = A_{is} - 1 = \frac{\theta_l(1-\theta_h)}{\theta_h(1-\theta_l)} - 1 = \frac{\theta_l-\theta_h}{\theta_h(1-\theta_l)} \quad . \tag{3.1}$$

With the partial cuts θ_a and θ_u for the auxiliary gas and the total UF$_6$, one correspondingly obtains for the separation factor of the mixture

$$A_m = \frac{\theta_a(1-\theta_u)}{\theta_u(1-\theta_a)} \quad . \tag{3.2}$$

The definitions given in (3.1,2) are equivalent to the usual equation defining the separation factor

$$A = \varepsilon_A + 1 = \frac{N'(1-N'')}{N''(1-N')} \quad , \tag{3.3}$$

which is based on the molar fractions N' and N'' of the component to be enriched in the light and the heavy fractions.

In the separation nozzle method, the technical expenditure and the corresponding costs are mainly due to the fact that large gas streams must be expanded in the separation stages and compressed again to the feed-pressure level. This expenditure is described by means of specific parameters which establish correlations between physical quantities typical of the process, e.g., the streams of material or the volume streams and the changes in state of such streams, on the one hand, and the separative work production (separative power)

$$\delta U = \theta_u(1-\theta_u)L_{uran} \, \varepsilon_A^2/2 \quad , \tag{3.4}$$

on the other hand /2,21,37-40/. L_{uran} is the uranium throughput through the separation element. In the following sections, the separative work will always be given in kilogram separative work units (SWU) and the separative power in kilogram separative work units per year (SWU/a).

The technical expenditure and, hence, the economics of the separation nozzle process, are characterized by the following specific parameters:

a) The *specific energy consumption* characterizes the compression work required to compress the gas expanded during the separation process and the technical expenditure for the electric system driving the compressors and for removing the compression heat.

b) The *specific suction volume* characterizes the sizes of the compressors, pipings, valves and tanks in a separation nozzle cascade.

c) The *specific slit length* characterizes the technical expenditure involved in manufacturing separation nozzle systems.

On the basis of the simplified assumption of an ideal isothermal compression of the mixture of auxiliary gas and UF_6 expanded in the nozzle and taking into account that the suction pressures p' and p" of the light and the heavy fractions may differ, one obtains for the specific energy consumption

$$E_s^{id} = L_m RT \ [\theta_m \ \ln(p_0/p') + (1-\theta_m) \ \ln(p_0/p")]/\delta U \quad , \tag{3.5}$$

where

L_m = mole throughput of the mixture of auxiliary gas and UF_6,
R = universal gas constant,
T = absolute temperature of the process gas,
θ_m = cut of the mixture of auxiliary gas and UF_6, and
p_0 = nozzle inlet pressure.

The values attainable in practice for the specific energy consumption depend, in addition, on the efficiency of the compressors, the electrical equipment, the cascade circuit, the pressure losses in the pipes and coolers, and the power consumption of the auxiliary systems; they are roughly 50% above the values given by (3.5). At the present state of development, the minimum specific power consumption of a large commercial separation nozzle plant is around 2800 kWh/SWU. This value corresponds to some 3% of the electricity generated with the enriched uranium in light water reactors; it is some 10% below the specific energy consumption of older gaseous diffusion plants and still 10% to 15% above that of the most recent gaseous diffusion plants /2,3/.

If the whole process gas mixture is assumed to be fed to the compressor at a suction pressure p' of the light fraction, the specific suction volume turns out to be

$$V_s^{id} = L_m RT/(p'\delta U) \quad . \tag{3.6}$$

When determining the specific suction volume for a technical plant, the mixing loss-es in the cascade and pressure losses outside the separation element must also be taken into account. Since, as is shown in Sect.4.2.3, the optimum operating pres-sure of the separation nozzle process is inversely proportional to the character-istic dimensions of the separation nozzle system, the minimum value of the specif-ic suction volume practically attainable is limited chiefly by the technical pos-sibilities of reducing the dimensions of the system. The specific suction volumes in commercial facilities attainable by the present fabrication techniques are around 5×10^5 m^3/SWU. Recent developments seem to indicate the possibility of another re-duction to values around 1×10^5 m^3/SWU. Compared with other enrichment techniques using pure UF$_6$ as the process gas, it must be taken into account as a matter of principle that the mixture of light auxiliary gas and UF$_6$ used in the separation nozzle process permits much higher flow rates to be attained than pure UF$_6$, be-cause of its low mean molecular weight and its correspondingly high sonic veloci-ty. Consequently, pipe cross sections and compressor sizes can be reduced consider-ably when operated on mixtures of auxiliary gas and UF$_6$ rather than on pure UF$_6$, if the same values are assumed in both cases for volume streams and pressure losses.

Let l be the length of a slit-type separation nozzle element and δU its separa-tive power; the specific slit length then turns out to be

$$l_s^{id} = 1/\delta U \quad . \tag{3.7}$$

Typical values of the specific slit length are between 1 and 3 m/(SWU/a), if the range of operation applicable for a commercial separation nozzle plant is used as a basis.

In addition to the specific parameters E_s^{id}, V_s^{id} and l_s^{id}, two other quantities are important when comparing with the gaseous diffusion process, namely the *equili-brium time* and the *number of stages* of the cascade. Also these quantities can be directly estimated from the physical and technical characteristics of the separa-tion stages and the characteristic parameters of the separation element for a given separation problem.

The equilibrium time of a cascade is the period in which an enrichment plant at-tains at the product and waste ends the desired molar fractions N_p and N_w, respec-tively, of the light uranium isotope ^{235}U if the whole plant in the beginning was filled with an isotope mixture of uniform molar fraction N_f. Since no material is removed from the plant over that period of time and the separative work performed is utilized only to build up the desired isotopic distribution in the cascade, it directly follows that the equilibrium time characterizes a technical expenditure of the enrichment process. The equilibrium time of the enrichment section of the cascade follows this formula:

$$\tau = \frac{2H}{\varepsilon_A^2} \cdot \frac{(N_p - 2N_p N_f + N_f) \ln(R_p/R_f)}{N_p - N_f} - 2 \tag{3.8}$$

with $R_p = N_p/(1-N_p)$, $R_f = N_f/(1-N_f)$.

The quantity H has the dimension of time and is frequently termed the stage transit time; it may be taken as the ratio between the materials inventory and the materials throughput of a separation stage. In the separation nozzle process, the average transit time is much shorter than in the gaseous diffusion process, which can be directly related to the low UF_6 content and the high sonic velocity of the mixture of auxiliary gas and UF_6, i.e., the low materials inventory and the high transport velocity of the materials. Since also the elementary effect ε_A is many times higher than in the gaseous diffusion process, the equilibrium time of a separation nozzle plant is more than one order of magnitude shorter than that of a gaseous diffusion plant. It is on the order of several hours if, starting from natural uranium, a product concentration of 3 mole % of ^{235}U and a waste concentration of 0.2 mole % of ^{235}U is used as a basis. The expenditure which depends on the equilibrium time, can be characterized in an equivalent way by the *specific uranium hold-up*,

$$G_s = G/\delta U \quad , \tag{3.9}$$

where G is the uranium inventory of the stage or of the plant and δU is the corresponding separative power. The specific parameter G_s, accordingly, characterizes the cost of the uranium inventory and the technical expenditure required to establish the desired isotopic distribution in the cascade. In the separation nozzle process, G_s is a few grams of uranium per SWU/a.

With respect to technical expenditure in an enrichment plant it is advantageous, as a matter of principle, to keep the number of separation stages of the cascade as small as possible. However, at the same time it must be ensured that the separation stages are run at uranium cuts at which the separative power of the separation elements is as high as possible and the cascade arrangement as simple as possible. In light of these constraints, cascade arrangements can be used for commercial-scale separation nozzle plants in which values of θ_u = 1/2, 1/3, 1/4 or 1/5 are set for the uranium cut of the separation stages. The number of stages Z of a separation nozzle cascade can then be described approximately by the relation

$$Z = \frac{1}{\theta_u \varepsilon_A} \cdot \ln \frac{N_p(1-N_w)}{N_w(1-N_p)} \quad , \tag{3.10}$$

where N_p and N_w are the ^{235}U molar fractions of the product removed from the plant and of the waste, respectively. The number of stages of a separation nozzle plant, depending on operating conditions and the type of separation nozzle, is between 200 and 500 if a product concentration of 3% of ^{235}U and a tails assay of 0.2% of ^{235}U is used as a basis. Consequently, it is a factor of 3 to 6 lower than the number of stages in a gaseous diffusion plant.

On the basis of the extensive planning work and the economic assessments associated with the commercialization of the separation nozzle process, the costs of uranium enrichment have been estimated for various plant concepts and broken down with respect to the various expenditures. The cost allocations obtained in this way can be used to assign weighting factors to the different specific parameters. These weighting factors reflect the importance of the different specific parameters in relation to the overall financial expenditure involved in the uranium enrichment process. It follows from such assessments that a cost fraction of 40% to 70% is due to the specific energy consumption, 10% to 25% to the specific suction volume, and 5% to 15% to the specific slit length. Another cost fraction which, among other items, characterizes the expenditure for instrumentation, auxiliary systems and plant buildings, is between 10% and 30%; this fraction is largely determined by the number of stages of the cascade. The large bandwidth of these cost factors is due to the fact that the fraction reflecting capital costs decreases with increasing plant size, while the operating cost fraction, which is mainly determined by electrical power consumption, rises by a corresponding margin. Moreover, very different power costs must be assumed, depending on the site of the plant.

4. Analysis of Separation Processes in the Separation Nozzle

4.1 Equilibrium Separation

In the separation nozzle, a mass element of the process gas mixture passes through a centrifugal field of a complicated spatial structure within a short period of time. Because of the limited flow time of the mixture in the centrifugal field and the spatial change of the centrifugal forces, no equilibrium distribution of the components of the mixture can be established at which the diffusion streams caused by pressure and concentration gradients would fully compensate each other at each point. Nevertheless, some fundamental aspects of the separation process can be described by the limit case of equilibrium separation, as will be shown below.

4.1.1 Distribution of Molecular Species in a Steady-State Cylindrical Flow

For simplification, an enclosed gas volume, as in a centrifuge, will be assumed to circulate in a cylindrical flow at a constant angular velocity ω; it is well known that, in this case, for each component i of the mixture a partial pressure distribution will be established of

$$p_i(r) = p_i(0) \exp[M_i \omega^2 r^2/(2kT)] \quad , \tag{4.1}$$

where r is the radial coordinate, M_i the mass of the molecule, k the Boltzmann constant, and T the absolute temperature. The quotient of the molar fraction ratios $N/(1-N)$, in the center $r = 0$, and at the periphery $r = r_0$ of the cylindrical flow, which follows from the partial pressures, is identical with the equilibrium separation factor $A*$ of a gas centrifuge /2,4,37/:

$$A* = \frac{N_1(0) (1-N_1(r_0))}{(1-N_1(0)) N_1(r_0)} = \exp[(M_h-M_1) \omega^2 r_0^2/(2kT)] \quad , \tag{4.2}$$

where N_1 and $N_h = 1-N_1$ are the molar fractions of the UF_6 molecules containing the light and the heavy uranium isotopes, respectively, in the total UF_6; M_1 and M_h are the respective molecular masses; and ωr_0 is the peripheral velocity of the cylindrical flow.

Introducing a flow parameter

$$S_i = v/c_i = \omega r/(2kT/M_i)^{1/2} \quad , \tag{4.3}$$

which indicates the ratio between the flow velocity v and the most probable thermal velocity c_i of the component i of the mixture - which is usually termed the speed ratio in gas dynamics - results in

$$A^* = \exp[(M_h-M_1)S_u^2(r_0)/M_u] \quad , \tag{4.4}$$

where M_u is the average molecular mass of UF_6.

While there is simple direct proportionality in the centrifuge of S_u and the peripheral velocity ωr_0 of the rotor, the UF_6 speed ratio in the separation nozzle is dependent in a complicated way on the operating conditions and the geometry of the separation system. Nevertheless, the increase in uranium isotope separation due to the acceleration by the light auxiliary gas can be explained in simple terms by determining the quantity A^* as a function of the peripheral Mach number and the UF_6 molar fraction N_u of the mixture of auxiliary gas and UF_6. At thermodynamic equilibrium, the Mach number of the mixture Ma, which characterizes the flow behavior of the mixture, is related to S_u by the relationship

$$S_u^2 = \frac{\gamma}{2} \cdot \frac{M_u}{M_m} Ma_m^2 = \frac{\gamma}{2} \cdot \frac{M_u}{N_u M_u + (1-N_u)M_a} \cdot Ma_m^2 \quad , \tag{4.5}$$

where γ is the ratio of specific heats of the mixture, M_a the molecular mass of the auxiliary gas, and M_m the average molecular mass of the mixture.

Figure 4.1 shows A^*-1 for the model case of a cylindrical flow with constant angular velocity as a function of the UF_6 molar fraction N_u for various Mach numbers Ma_m of an H_2/UF_6 mixture. The relative difference between the isotopic ratios at the periphery and in the center of the cylindrical flow is seen to increase by more than one order of magnitude if a mixture of H_2/UF_6 with $N_u = 0.02$ to 0.05, which is typical of the separation nozzle process, is employed instead of pure UF_6 ($N_u = 1$) at a given Mach number Ma_m. This increase in the relative difference A^*-1 is obviously due to the fact that the UF_6 speed ratio S_u increases correspondingly with a reduction in mean molecular weight of the mixture at a given Mach number Ma_m of the mixture.

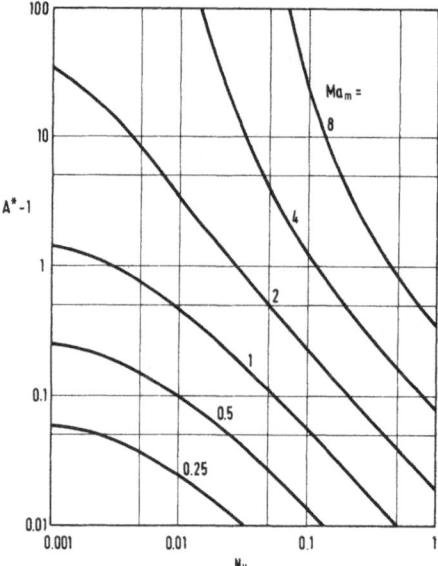

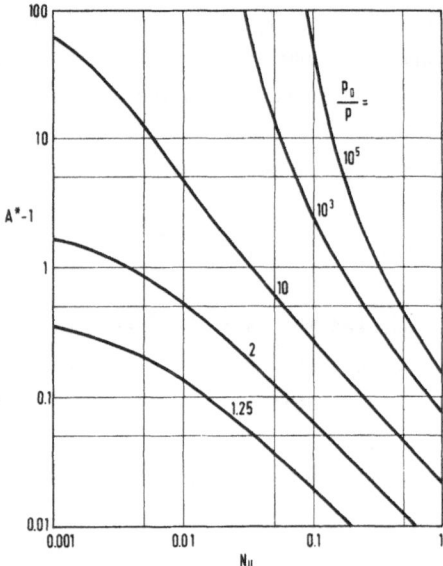

Fig.4.1. Influence of the peripheral Mach number $Ma_m(r_0)$ and the UF_6 molar fraction $N_u(r_0)$ upon the relative difference A*-1 between the isotopic ratios in the center and at the periphery of an isothermal cylindrical flow of an H_2/UF_6 mixture

Fig.4.2. Influence of the expansion ratio p_0/p and the UF_6 molar fraction N_u upon the relative difference A*-1 between the isotopic ratios in the center and at the periphery of a cylindrical flow of an H_2/UF_6 mixture

Under the simplifying assumption of the mixture of auxiliary gas and UF_6 being accelerated to a certain peripheral Mach number Ma_m by isentropic expansion it is possible, within the framework of the model described here, to establish a relationship between the expansion ratio p_0/p of the flow and the relative variation of the isotopic ratio in the flow field. From the corresponding model calculations shown in Fig.4.2 it is evident that, under these conditions, a mixture of H_2/UF_6 with 4 mole % of UF_6 would reach a difference of 15% in the isotopic ratios at the periphery and in the center of the cylindrical flow at an expansion ratio of $p_0/p = 2$. With pure UF_6, an expansion by a factor of 10^5 would be necessary for the same change in the isotopic ratio, which would require extremely high volume flows to be extracted from the separation element and would lead to correspondingly infavorable values for the specific suction volume. The technical expenditure for the isotope separation process can thus be reduced essentially by the use of a light auxiliary gas.

4.1.2 Bifractional Splitting

In a gas centrifuge, the gas stream circulating with the angular velocity of the rotor is always large in comparison with the feed and extraction streams. If a simple cocurrent device is considered, the elementary effect of isotope separation is given by the isotopic ratios at the axis and at the periphery of the rotor, where the small streams of the light and heavy fractions are withdrawn (4.2). In contrast to a gas centrifuge, the total gas stream passing through the centrifugal field is split into two fractions in a separation nozzle, i.e., the circulating gas stream is identical with the feed stream and, correspondingly, with the sum of the extracted streams. Accordingly, the elementary effect of isotope separation of a separation nozzle depends on the comparatively small relative difference between the average isotopic ratios of the two parts of the cylindrical gas stream, while in a cocurrent centrifuge the separation effect is determined by the large relative difference between the local isotopic ratios in the center and at the periphery of the circulating stream.

The elementary effect of isotope separation between the two fractions of the nozzle can be calculated from the partial cuts θ_i of the light and the heavy species of molecules (3.1). For a flow with concentric stream lines which is split into two partial streams at the point r, the θ_i values can be calculated directly by radial integration over the flux profile $j_i(r)$:

$$\theta_i(r) = \int_0^r j_i(r)dr / \int_0^{r_0} j_i(r)dr \quad . \tag{4.6}$$

In the following considerations, as in the previous section, a cylindrical flow with constant angular velocity ω and spatially constant temperature is assumed, which is in a state of diffusional equilibrium. With the gas equation

$$p_i = \nu_i kT \tag{4.7}$$

and (4.1) one obtains for the radial development of the flux of the component i

$$j_i(r) = \omega r \nu_i(0) \exp[M_i \omega^2 r^2 / (2kT)] \quad , \tag{4.8}$$

where ν_i is the number density and M_i the molecular mass.

Substituting (4.8) in (4.6) yields

$$\theta_i(r) = \frac{\exp[(r/r_0)^2 s_i^2(r_0)] - 1}{\exp[s_i^2(r_0)] - 1} \quad , \tag{4.9}$$

18

where $S_i(r_0)$ is the speed ratio of the component i at the periphery ($r = r_0$) of the cylindrical flow.

Because of the small relative difference in the molecular masses, the partial cuts θ_1 and θ_h of the light and the heavy species of molecules differ only slightly and correspond approximately to the UF_6 cut θ_u. Therefore, the elementary effect of isotope separation for a cylindrical flow in the state of diffusional equilibrium can be indicated directly as a function of the cut θ_u and the UF_6 speed ratio S_u /35/. This dependence of the equilibrium separation effect ε_A^* on S_u and θ_u is shown in Fig.4.3. It is seen that ε_A^*, at a given UF_6 cut, initially rises steeply with the UF_6 speed ratio and then converges against a limit. For $\theta_u = 1$, this limit is equal to the relative mass difference $(M_h - M_1)/M_u$; the high speed ratio limit of ε_A^* rises continuously with decreasing θ_u.

The existence of a bound of the equilibrium separation effect ε_A^* which is dependent only on the UF_6 cut can be explained by the fact that at $S_u \gtrsim 4$, the molar fraction gradients of the isotopes increase with S_u in precisely the same proportion as

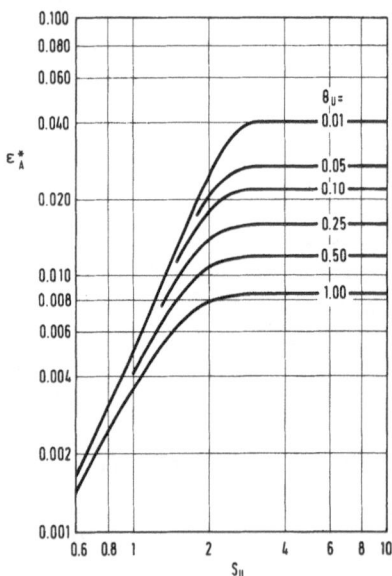

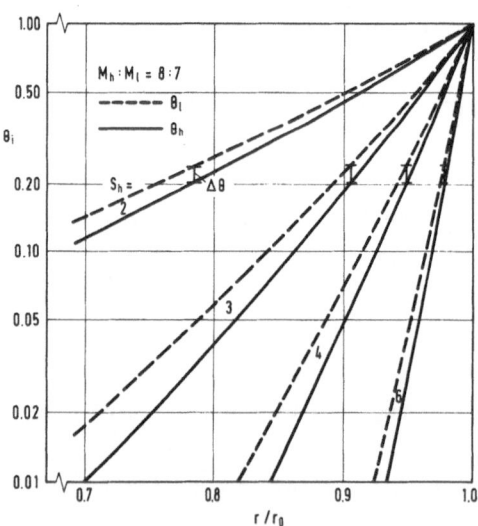

Fig.4.3. Influence of the UF_6 speed ratio S_u and the UF_6 cut θ_u upon the equilibrium separation effect ε_A^* under conditions of bifractional splitting of an isothermal cylindrical flow at a constant angular velocity

Fig.4.4. Radial plot of the partial cuts θ_1 and θ_h of the light and the heavy components of a gaseous mixture in a cylindrical flow with constant angular velocity in diffusion equilibrium (S_h = speed ratio of the heavy component at the periphery of the centrifugal field; ratio of molecular weights, $M_h/M_1 = 8/7$)

the UF_6 is forced to the periphery of the centrifugal field. This is illustrated
by Fig.4.4, in which the partial cuts θ_l and θ_h of the light and the heavy components of a model gas mixture are plotted as a function of the radius r for a cylindrical flow with constant angular velocity in a state of diffusion equilibrium. The
molecular weights of the light and the heavy components were assumed to behave as
7:8 so that the differences between θ_l and θ_h could be resolved graphically in the
diagram. The speed ratios S_h indicated in Fig.4.4 apply to the heavy component at
the periphery of the centrifugal field, $r/r_0 = 1$. It is seen that at a given radius,
the relative difference in partial cuts $\ln \Delta \theta$ and, hence, the separation effect increases with increasing speed ratio. However, if one starts at a given value of θ_h,
the relative difference in partial cuts for $S_h \gtrsim 4$ is practically independent of the
speed ratio; a further increase in S_h leads only to an increasing accumulation of
the model gas mixture at the periphery of the centrifugal field, but not to a further increase in the separation effect at a given cut.

If it is taken into account that the speed ratios of the components of the mixture in thermodynamic equilibrium behave like the square roots of the molecular
masses, it follows from (4.9) for high values of S_u that

$$\theta_l = \exp\left[(M_l/M_u)S_u^2 \, (r^2 - r_0^2)/r_0^2\right] = \theta_u^{M_l/M_u} \tag{4.10}$$

and

$$\theta_h = \theta_u^{M_h/M_u} \quad . \tag{4.11}$$

Substitution of the partial cuts θ_l and θ_h in (3.1) yields the upper bound of the
equilibrium separation effect for small relative differences in masses /35/

$$\varepsilon_{A,\infty}^{\star} = \frac{M_l - M_h}{M_u} \cdot \frac{\ln \theta_u}{1 - \theta_u} \quad . \tag{4.12}$$

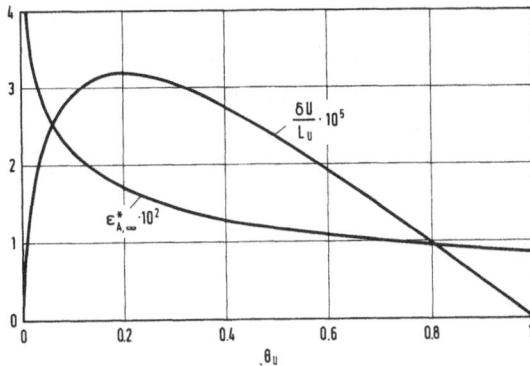

Fig.4.5. Upper bound of the equilib-
rium separation effect $\varepsilon_{A,\infty}^{\star}$, and the
corresponding normalized separative
work output $\delta U(\varepsilon_{A,\infty}^{\star})/L_u$ as a function
of the UF_6 cut θ_u

Figure 4.5 shows $\varepsilon^*_{A,\infty}$ and the separative power normalized to the UF$_6$ throughput $\delta U(\varepsilon^*_{A,\infty})/L_u$ which formally results from (3.4) and (4.12), plotted as a function of the UF$_6$ cut θ_u. Since the relative difference in partial cuts θ_l and θ_h and the relative difference in local molar fractions of the light and heavy isotopes rise continuously from the periphery to the center of the circular flow, ε^*_A and, correspondingly, $\varepsilon^*_{A,\infty}$ will increase with decreasing θ_u. A relationship analogous to (4.12) is obtained for the change in molar fraction of the residue in Rayleigh distillation /37/. Accordingly, (4.12) can be derived under the condition that the relative change dn/n of the ^{235}U molar fraction within a control volume defined by a cylindrical surface around the axis of the centrifugal field is equal to the decrease dq/q of the UF$_6$ inventory in this control volume, multiplied by the relative mass difference $\Delta M/M$. For the function indicated in (4.12), the maximum separative power is reached at $\theta_u = 0.2$, i.e., at a highly asymmetrical splitting of the UF$_6$ stream L_u.

Besides the splitting of a separation nozzle flow into two fractions, in theory splitting into three or more fractions is also possible. If the equilibrium separation effect is calculated on the basis of the mean molar fractions in the two partial streams spaced the largest distance apart, the values resulting for ε^*_A will increase if these two partial streams are reduced relative to the total flow rate L_u. For the model case of cylindrical flow at a constant angular velocity, $\varepsilon^*_A + 1$ in a trifractionation converges directly into the equilibrium separation factor of the centrifuge as defined in (4.2), if the partial streams extracted at the periphery and in the center become small relative to the total stream.

When splitting the isotopic distribution of a centrifugal flow into two fractions, relatively large mixing losses occur because within each fraction the gradients in the isotopic ratios generated by the centrifugal field are equalized downstream from the splitting point. These mixing losses can be reduced by splitting the flow into more than two fractions, thus allowing a higher separative work output to be achieved on the whole. However, this increase in separative work output can only be exploited if no additional mixing losses occur as a result of the interconnection of separation elements in a cascade. For this reason, only an arrangement for trifractionation is of practical importance in which the isotopic composition of the intermediate fraction corresponds to that of the feed gas, thus allowing it to be combined with that gas without further mixing losses (Sect.8.2).

4.2 Uranium Isotope Separation in the Auxiliary Gas

In equilibrium separation, the light auxiliary gas plays a role only insofar as it allows the UF_6 to be accelerated to high speed ratios even at low expansion ratios of the auxiliary gas/UF_6 mixture. At a given UF_6 speed ratio, the distribution of the isotopes is not influenced by the auxiliary gas in a state of diffusion equilibrium. However, if one considers the spatial development of uranium isotope separation in a curved flow, one must take into account not only the effect of accelerating the UF_6, but also the effect of the auxiliary gas upon diffusion processes in the flow field /35/.

A precise description of this ternary diffusion process, in which the gas kinetic collisions of the three constituents of the mixture must be considered, cannot be given by the theoretical methods now available. On the one hand, this is due to the fact that the large mass difference between UF_6 and the auxiliary gas, the high velocity gradients in the flow field, and the relatively small number of gas kinetic collisions highly disturb the molecular velocity distribution over broad ranges of the separation nozzle, i.e., the local velocity distribution of the molecules cannot even be approximately described by a Maxwellian distribution. On the other hand, it must be taken into account that the pronounced separation between UF_6 and the auxiliary gas is associated with a marked transport of energy and momentum in the flow field, which results in strong coupling of the flow and diffusion processes (cf., e.g., /33/).

For the above reasons the flow was always presumed to have a simplified velocity field in the theoretical studies conducted so far on separation processes. Moreover, the local velocity distribution of the molecules was usually assumed to correspond approximately to a Maxwellian distribution, allowing diffusion processes to be treated within the framework of the Chapman-Enskog approximation of the Boltzmann equation.[3] The analysis of uranium isotope separation in the separation nozzle which can be carried out subject to these constraints is explained in more detail below.

---

[3] To elucidate the influence which major disturbances in molecular velocity distribution have on isotope separation, Monte Carlo calculations were recently started /41,42/. No results are as yet available for direct comparison with the calculations based on the Chapman-Enskog approximation. Theoretical studies in which approaches other than the Chapman-Enskog approximation were used to solve the Boltzmann equation (moment method, multifluid model) have not yet yielded tangible results /42-44/.

4.2.1 Ternary Diffusion Processes

The process of isotope separation in the separation nozzle is determined by the diffusion of the isotopes, which permeate the streamlines or, strictly speaking, the molar stream surfaces of the isotopic mixture, when the flow is deflected. A UF_6 molar stream surface splits the UF_6 stream entering the separation nozzle L_u into two partial streams, $\theta_u L_u$ and $(1-\theta_u)L_u$, thus representing a surface of constant UF_6 cut. As a result of the curvature of the stream surfaces, a component in the pressure gradient is developed which is normal to the direction of flow, and pressure diffusion causes the light isotope to be transported to the concave side and the heavy isotope to the convex side of a UF_6 molar stream surface. In the course of deflection, an increasing concentration gradient is established at the UF_6 molar stream surface, and the concentration diffusion streams directed opposite to the pressure diffusion streams increase, thus making net transport smaller and smaller with increasing angle of deflection. Since the UF_6 is concentrated increasingly at the periphery of the centrifugal field with increasing angle of deflection, the molar fraction gradients of the isotopes can become so high in this region that the remixing concentration diffusion streams exceed the separating pressure diffusion streams. This remixing effect is additionally favored by the fact that the flow velocity and the radial pressure gradient at the periphery of the centrifugal field are smaller than in the central region of the flow field, due to viscous effects and the lesser expansion in these flow regions.

The differences in the average isotopic ratios on both sides of the UF_6 molar stream surface are best characterized by the elementary effect ε_A of isotope separation which, according to (3.1), can be determined by calculating the partial cuts. The differential equation for the spatial development of the partial cut θ_i is obtained by regarding the transport of the isotope i through an elementary area of a stream surface of the auxiliary gas/UF_6 mixture. If it is assumed for simplification that the molar stream surfaces of the mixture correspond to concentric cylinders in the separation nozzle, it holds that

$$\partial\theta_i = \frac{1}{L_i} j_i r\partial\phi \quad , \tag{4.13}$$

where L_i is the throughput of the isotope i per unit length of the nozzle, j_i the radial component of the flux, and $r\partial\phi$ the elementary area. The flux j_i is composed of two components resulting from the motion of the isotopic species relative to UF_6 and the motion of UF_6 relative to the mixture of the auxiliary gas and UF_6.

The calculation of j_i is based on the binary diffusion equation

$$\underline{j}_i = \nu_i \underline{w}_i = -\nu D\left[grad\ N_i - (\Delta M/\overline{M})N_i(1-N_i)(1/p)grad\ p\right]^{\cdot} \quad , \tag{4.14}$$

where ν_i is the number density of the component i, $\underline{w}_i$ the diffusion velocity, ν the total number density, D the diffusion coefficient, and p the static pressure. Equation (4.14) contains only the terms describing concentration and pressure diffusions. The influence of thermal diffusion can be neglected because of the low temperature gradients in the separation nozzle (Sect.5.3.2 and /45/) and because of the small thermal diffusion factor /46/.

In the limiting case of small molar fractions n of the light isotope in UF_6 ($n \ll 1$, $M_h \cong M_u$), the relative diffusion of UF_6 and auxiliary gas is not influenced greatly by the light isotope. The diffusion of the light isotope in the mixture can thus be treated like diffusion in a uniform gas, the diffusion coefficient following approximately from the relation /47/

$$1/D_{a,h}^1 = N_h/D_{1,h} + (1-N_h)/D_{a,u} = 1/D_T \quad . \tag{4.15}$$

In this relation, $D_{1,h}$ and $D_{a,u}$ are the binary diffusion coefficients of the isotopic mixture $^{235}UF_6/^{238}UF_6$ and of the auxiliary gas/UF_6 mixture, respectively.[4] D_T is usually termed the ternary diffusion coefficient.

If the relative pressure gradient in (4.14) is replaced by the radial term of the Eulerian equation

$$(1/p)\partial p/\partial r = \overline{M}v^2/(kTr) \tag{4.16}$$

(v = flow velocity of the mixture = azimuthal velocity), the differential equation for the spatial dependence of the partial cut of the light isotope will be found to be

$$\frac{\partial \theta_1}{\partial \phi} = \frac{N_u}{N_u^0} \cdot \frac{\nu D_T}{L_m} \left[\frac{\partial(n/n_0)}{r\partial r} + \frac{n}{n_0} \cdot \frac{(M_u-M_1)v^2}{kT} \right] + \frac{n}{n_0} \frac{\partial \theta_u}{\partial \phi} \quad , \tag{4.17}$$

where N_u^0 is the average molar fraction of the UF_6 in the mixture and n_0 the average molar fraction of the light isotope in UF_6 /35/.

By analogy with (4.17), the relation

$$\frac{\partial \theta_u}{\partial \phi} = \frac{\nu D_m}{L_m} \left[\frac{\partial(N_u/N_u^0)}{r\partial r} - \frac{N_u}{N_u^0} \cdot \frac{(M_u-M_m)v^2}{kT} \right] + \frac{N_u}{N_u^0} \cdot \frac{\partial \theta_m}{\partial \phi} \tag{4.18}$$

---

[4] In the model calculations described below, the values of $\nu D_{1,h} = 0.4 \times 10^{18}$ cm^{-1}s^{-1} /48/ and $\nu D_{a,u} = 8.5 \times 10^{18}$ cm^{-1}s^{-1} /49/ obtained at T = 293 K were used.

applies to the differential change in the uranium cut, $\partial\theta_u/\partial\phi$.

The partial differential equations (4.17,18) were solved numerically in /35/ for H_2/UF_6 mixtures with different UF_6 molar fractions N_u^0. In accordance with flow studies conducted on the standard separation nozzle (Fig.2.1) a parabolic velocity profile was taken as a basis which, for simplification, was assumed to be independent of the deflection angle (cf. Sect.5.3.2 and /50/). The maximum velocity given for an H_2/UF_6 mixture with 5 mole % UF_6 was 450 m/s, and a constant temperature of T = 293 K was assumed throughout the entire flow field. Conversion to the velocities at other UF_6 molar fractions N_u^0 was then achieved by multiplication by the square root of the reciprocal ratio of the mean molecular weights.

Figure 4.6 shows some typical results of the azimuthal dependence of the elementary effect ε_A of isotope separation and the separation factor of the mixture A_m, between H_2 and UF_6 for a constant uranium cut θ_u = 1/3. It is seen that ε_A passes through a clear maximum over the angle of deflection at the lowest UF_6 molar fraction (1.5 mole %), the maximum of ε_A of 1.8% clearly exceeding the upper limit of the equilibrium separation effect $[(\varepsilon_{A\infty}^{\star}(\theta_u=1/3)) = 1.41\%, \text{cf. Fig.4.5 and}$ (4.12)]. With increasing UF_6 molar fraction ε_A decreases markedly, and the maximum is less pronounced and is reached at higher angles of deflection. Unlike ε_A, the separation factor of the mixture A_m does not pass through a maximum over the angle of deflection, but asymptotically approaches a limit. The asymptotic value of A_m decreases with increasing UF_6 molar fraction N_u^0 because the flow velocity decreases with increasing molecular weight.

The decrease of ε_A with increasing UF_6 molar fraction and the slower rise of ε_A over the angle of deflection can, according to (4.17), be explained primarily by

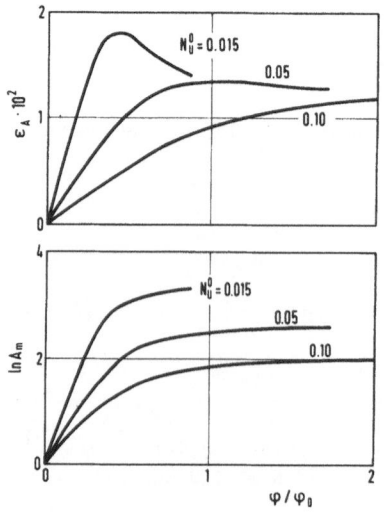

Fig.4.6. Results of model calculations on the azimuthal dependence of the elementary effect ε_A of isotope separation and the separation factor A_m of the mixture for H_2/UF_6 mixtures with various UF_6 molar fractions N_u^0. The calculations were based on a parabolic velocity profile independent of the angle of deflection ϕ

the reduction of the flow velocity. Moreover, the ratio of the ternary to the binary diffusion coefficient D_T/D_m decreases, which also reduces the separation of isotopes in the auxiliary gas/UF_6 mixture. The value of ε_A arising at infinite angles of deflection corresponds to the equilibrium separation effect in case of bifractional splitting of the flow. This asymptotic value depends only on the absolute value and the radial profile of the UF_6 speed ratio, but no longer on D_T and D_m (Sect.4.1.2).

The enhancement of the elementary effect during the transition to diffusion equilibrium, which follows from the model calculations shown in Fig.4.6, is in good qualitative agreement with corresponding experiments performed on H_2/UF_6 mixtures. However, it is difficult to give a simple idea of this phenomenon on the basis of the formalism developed so far. Therefore a simplified model of isotope separation in the auxiliary gas is treated below. It is based on the assumption that UF_6 is present only in trace amounts in the auxiliary gas, and allows the transient enhancement of the elementary separation effect to be described relatively simply /51-53/.

4.2.2 Isotope Separation in Highly Dilute UF_6

At very low UF_6 molar fractions in the auxiliary gas, the collisions between the UF_6 molecules can be neglected relative to the collisions with the auxiliary gas molecules, and the ternary diffusion equations change into two uncoupled binary diffusion equations. If one takes into account that only the diffusion streams normal to the stream surfaces contribute to the shift in concentration, substitution of the diffusion flux (4.14) into the continuity equation of the component i gives the following differential equation for the spatial change in the molar fraction N_i:

$$\frac{v}{r}\frac{\partial N_i}{\partial \phi} = \frac{D}{r} \cdot \frac{\partial}{\partial r}\left[r\left(\frac{\partial N_i}{\partial r} - \frac{M_i - M_a}{kT}N_i\frac{v^2}{r}\right)\right] \quad , \tag{4.19}$$

where D is the binary diffusion coefficient of the mixture of auxiliary gas/UF_6 and v is the flow velocity of the mixture.

If the radial coordinate r is normalized to a characteristic dimension of the separation nozzle, e.g., the radius of curvature r_0 of the deflection wall, and the velocity v is normalized to a reference velocity v_0 (4.19) can then be written in the form

$$\frac{\partial N_i}{\partial \hat{\phi}} = \frac{1}{\hat{v}}\frac{\partial}{\partial \hat{r}}\left(\hat{r}\frac{\partial N_i}{\partial \hat{r}} - 2\beta_i N_i\right) \quad , \tag{4.20}$$

where $\hat{v}$, $\hat{r}$ and

$$\hat{\phi} = \phi D/(r_0 v_0) \tag{4.21}$$

represent normalized quantities. The dimensionless quantity

$$\beta_i = (M_i - M_a)v^2/(2kT) = S_i^2 - S_a^2 \tag{4.22}$$

depends only on the speed ratios S_i of the isotopes and the speed ratio S_a of the auxiliary gas.

The partial differential equation (4.20) can be solved by separating the variables for cylindrically symmetrical flow fields /51,52,54/. The solution can be represented by the eigenfunctions $P_{ik}(\hat{r})$ and the corresponding eigenvalues λ_{ik} as

$$N_i(\hat{r},\hat{\phi}) = N_i^*(\hat{r}) + \sum_{k=1}^{\infty} C_{ik} \exp(-\lambda_{ik}\hat{\phi}) \ P_{ik}(\hat{r}) \quad , \tag{4.23}$$

where $N_i^*(\hat{r})$ indicates the radial distribution of the isotope i in the diffusion equilibrium.

The spatial development of the molar fraction according to (4.23) obviously characterizes a transient process in which an initial distribution $N_i(\hat{r},\hat{\phi}=0)$ asymptotically changes into the equilibrium distribution $N_i^*(\hat{r})$, with an increasing normalized angle of deflection $\hat{\phi}$. The eigenvalues λ_{ik} increase with the quantity β_i defined in (4.22), as was shown in /51/ for a cylindrical flow with constant angular velocity. Since under conditions of local thermal equilibrium, the speed ratio of the heavy isotope is higher than that of the light isotope and $\beta_h > \beta_l$, the heavy isotope attains its equilibrium distribution more quickly than the light isotope. As a consequence, during the transient process a condition is passed in which the heavy isotope is already concentrated at the periphery of the centrifugal field, while the light isotope is still distributed over a wider range.

In the part of the flow situated on the concave side of a UF_6 stream surface, the molar fraction n_l' of the light isotope in the UF_6 initially rises during the transient process while, correspondingly, the molar fraction n_h'' of the heavy isotope increases on the convex side. Since in transition to equilibrium distribution, the UF_6 is concentrated in an increasingly narrow layer at the periphery of the centrifugal field, the molar fraction gradients and, hence, the concentration diffusion streams also increase. Because of the more rapid equilibrium adjustment in the heavier isotope, this rise occurs earlier in it than in the light isotope; consequently, the difference in diffusion velocities of the light and the heavy isotopes will change signs at a certain angle of deflection and, correspondingly, at a certain radial position of the UF_6 molar stream surface. As a result, the relative differences in isotopic ratios on both sides of the UF_6 molar stream surface will decrease again, i.e., the elementary effect of isotope separation at a given uranium cut will decrease again and converge asymptotically into the equilibrium value ε_A^*. Since the transition into diffusion equilibrium in a ternary mixture is always connected with remixing, equilibrium separation, unlike the condition in a binary mixture, is not identical with maximum separation of the isotopes.

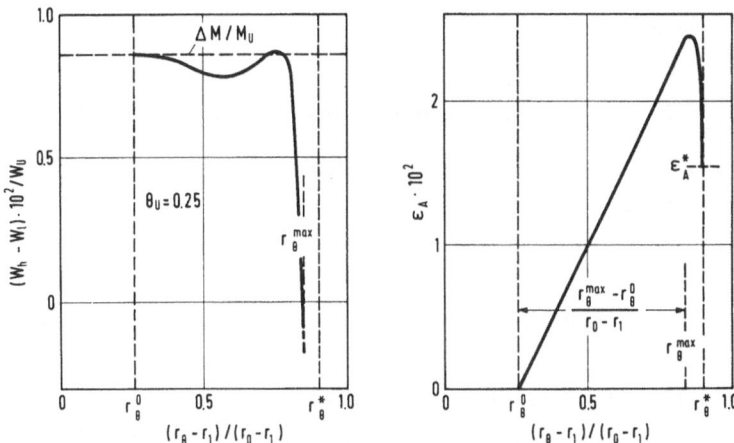

Fig.4.7. Relative difference, $(w_h-w_l)/w_u$, of the isotopic diffusion velocities and elementary effect ε_A of isotope separation versus radial position r_θ of the UF_6 molar stream surface $\theta_u = 0.25$. The model calculation was based on an azimuthally and radially constant velocity and temperature corresponding to $S_u = 4$ and a negligibly small UF_6 molar fraction; r_0 and r_1 denote the outer and the inner boundaries, respectively, of the curved flow

This situation is shown in Fig.4.7, in which the relative difference $(w_h-w_l)/(w_u)$ in diffusion velocities of the isotopes, and the elementary effect ε_A of isotope separation for a constant uranium cut $\theta_u = 0.25$ are plotted as a function of the spatial position r_θ of the corresponding UF_6 stream surface. In the model calculation it was assumed for simplification that the flow velocity is constant both in the radial and the azimuthal directions; for the UF_6 speed ratio a value of $S_u = 4$ was assumed. It is seen that ε_A initially rises with increasing r_θ, passes through a maximum at r_θ^{max} and, when approaching the equilibrium position r_θ^*, drops to the equilibrium separation effect ε_A^*. At the r_θ^{max} position, the difference between the diffusion velocities of the light and the heavy isotopes changes sign, while the diffusion velocity of UF_6 in the auxiliary gas decreases monotonically and vanishes at r_θ^*. Accordingly, in contrast to ε_A, the separation factor of the mixture rises continuously with increasing r_θ, attaining its maximum asymptotically at the equilibrium position r_θ^*.

The dependence on the UF_6 cut θ_u of the maximum elementary effect of isotope separation, $\varepsilon_A^{max} = \varepsilon_A(r_\theta^{max})$, except for a constant factor, largely corresponds to the relationship derived from (4.12) for the upper limit of equilibrium separation. This can be explained by the fact that the relative increase dn/n in the ^{235}U molar fraction in UF_6 on the concave side of a molar stream surface of the auxiliary gas/UF_6 mixture is proportional to the relative decrease of the UF_6 molar fraction in this part of the flow. This behavior is analogous to the Rayleigh distillation process as indicated already in Sect.4.1.2.

4.2.3 Influence on Diffusion Processes of the Knudsen Number and the Speed Ratio

The spatial pattern of the molar fractions of the isotopes as described by (4.23) remains unchanged at a given speed ratio, if the nondimensional quantity $D/(r_0 v_0)$, which corresponds to the reciprocal value of the Peclet number used in mass transfer processes, remains constant. At a constant reference velocity v_0 it is possible to reduce $D/(r_0 v_0)$ directly to the Knudsen number Kn of the flow, i.e., to the ratio of the mean free path and a characteristic dimension r_0 of the separation nozzle. Since the product of the total number density ν and the diffusion coefficient D may be regarded as constant under the conditions prevailing in the separation nozzle, the relation

$$\hat{\phi}/\phi = \nu D/(\nu r_0 v_0) = const\cdot\lambda/r_0 = const\cdot Kn \tag{4.24}$$

follows from (4.21). In a cylindrical flow with given azimuthally constant values of velocity and temperature, one thus always obtains the same relationship between the elementary effect ε_A of isotope separation and the UF_6 cut θ_u if the product Kn·ϕ is kept constant. The Knudsen number is inversely proportional to the product of the operating pressure p_0 and a characteristic dimension r_0 of the separation nozzle. Consequently, the relation $\varepsilon_A(\theta_u)$ remains unchanged for a given angle of deflection, if an increase in the operating pressure is counterbalanced by a corresponding decrease in the characteristic dimensions of the separation nozzle. If the operating pressure and the angle of deflection are changed in the same proportions at a given value of r_0, the relation $\varepsilon_A(\theta_u)$ still remains unchanged.

The eigenvalues λ_{ik}, which determine the transient process, rise roughly proportional to the square of S_u at high UF_6 speed ratios S_u, as was shown in /51/ by the example of a cylindrical flow with constant angular velocity. Since the transit time of the gas through a certain angular sector decreases only with the reciprocal value of the velocity and, at a constant temperature, only with the reciprocal value of S_u, the separation process proceeds more quickly over a given angle of deflection if the

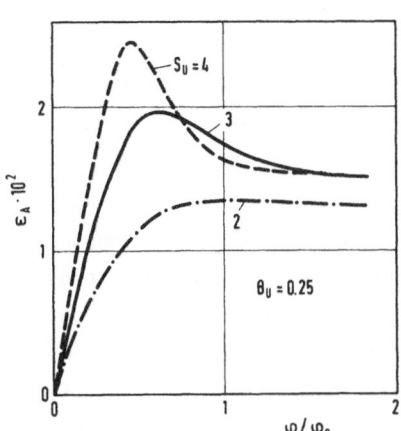

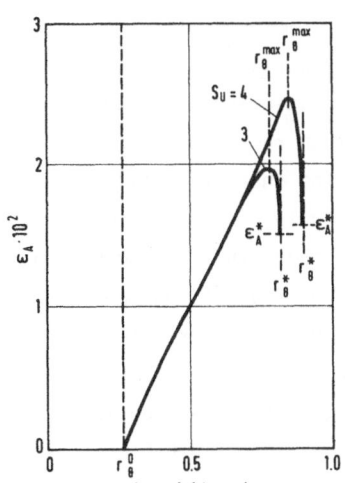

Fig.4.8. Influence of the UF₆ speed ratio S_u on the azimuthal development of the elementary effect ε_A of isotope separation for the UF₆ stream surface $\theta_u = 0.25$ $[v(r,\phi) =$ const, $N_u \rightarrow 0]$

Fig.4.9. Influence of the UF₆ speed ratio S_u on the radial development of the elementary effect ε_A of isotope separation for the UF₆ stream surface $\theta_u = 0.25$ $[v(r,\phi) =$ const, $N_u \rightarrow 0]$

speed ratio increases (Fig.4.8). This finding can be substantiated qualitatively by indicating that the separating pressure diffusion stream rises proportional to S_u^2, while, in an isothermal flow, the azimuthal transport of UF₆ increases only proportional to S_u. If one assumes that in optimum operation of a separation nozzle the maximum of the elementary effect of isotope separation is to be attained just where the flow is split by the skimmer, it follows that an increase in the speed ratio must be associated with a reduction in the Knudsen number of the flow. For practical purposes, this means that the optimum operating pressure of a separation nozzle rises, if the expansion ratio is increased or the UF₆ molar fraction of the process gas is reduced.

If the UF₆ speed ratio S_u is increased, the equilibrium position r_θ^{max} of a UF₆ molar stream surface and, consequently, r_θ^{max} will be shifted towards the periphery of the centrifugal field. The resultant increase in the relative diffusion path, i.e., the increase in the relative displacement between UF₆ and light auxiliary gas, results in an increase in the maximum elementary effect ε_A^{max} as can be taken from the results of corresponding model calculations shown in Fig.4.9. Since the radial diffusion path of a UF₆ stream surface is limited by the deflection wall, ε_A^{max} cannot be increased arbitrarily by raising the UF₆ speed ratio, but converges against an upper limit.

Some direct conclusions about the specific process parameters (Sect.3) can be derived from the conditions elaborated above. At a constant expansion ratio, the specific energy consumption remains unchanged if the characteristic dimensions are varied inversely proportional to the operating pressure. This is due to the fact that the UF_6 speed ratio and the Knudsen number and, hence, the elementary effect of isotope separation remain unchanged as well as the mass throughput and the compression work. If the characteristic dimensions of the separation nozzle are decreased or if the angle of deflection is increased, the specific suction volume is reduced because of the resulting increase in the optimum operating pressure. At a constant Knudsen number and a constant expansion ratio, the specific slit length is independent of the characteristic dimensions, because the UF_6 throughput through the separation element and the elementary effect of isotope separation remain unchanged. If the angle of deflection is increased, the specific slit length decreases, because the optimum Knudsen number becomes smaller and, hence, the optimum UF_6 flow through the nozzle becomes larger. With increasing expansion ratio, the speed ratio rises and the optimum Knudsen number decreases, which results in a higher elementary effect of isotope separation *and* a higher UF_6 throughput. Therefore, an increase in the expansion ratio causes the specific slit length to decrease continuously to a limit which, inter alia, is dependent on the geometry of the separation nozzle.

4.2.4 Influence on Isotope Separation of the Spatial Structure of the Flow Field

The model concept described in Sect.4.2.2 on the development of isotope separation in the separation nozzle clearly shows that with the uranium cut remaining constant the elementary effect of isotope separation increases the greater the shift of the corresponding UF_6 stream surface relative to the auxiliary gas. Consequently, it is advantageous for isotope separation if the radial position of a UF_6 stream surface at the beginning of deflection differs largely from its radial position at the skimmer. In order to meet this requirement, the UF_6 flux j_u at the nozzle inlet should be as high as possible in small radii and as low as possible in large radii /52/. Consequently, the radius of curvature r_1 of the inner boundary of the flow should then be as small as possible, i.e., the ratio of the width r_0-r_1 of the curved flow to the radius of curvature r_0 of the deflection wall should be as high as possible. It is also evident that a flow profile with radially decreasing velocity leads to a more advantageous starting position of the UF_6 stream surfaces than a flow profile with radially increasing velocity, if one starts at a subsonic flow of the mixture, i.e., a flow in which the flux increases with the velocity.

The higher the centrifugal field of the flow, the more a UF_6 stream surface can be shifted towards the periphery of the centrifugal field during deflection (Fig. 4.9). However, the separation effect between the light and heavy fractions is not determined by the mean strength of the centrifugal field during the whole deflection process, but by the field strength in the flow region right upstream from the skimmer. In the flow regions further upstream, an increase in centrifugal forces results only in an acceleration of the separation process, but has no direct impact on the separation effect to be achieved at the skimmer /52/.

4.2.5 Experimental Verification of the Transient Enhancement in Isotope Separation

The model concepts of the process of isotope separation in a curved flow of a mixture of auxiliary gas and UF_6 can be verified by a number of experiments, some of which will be described in more detail below.

Figure 4.10 shows the results of separation experiments performed in a standard separation nozzle on a mixture of H_2/UF_6 with 1.6 mole % of UF_6 at the inlet pressure optimal for separation. The gas mixture was expanded by a factor of 8, and the separation effect was determined at various radial positions of the skimmer, i.e., for various UF_6 cuts. It is seen that the values measured for the separation effect under these conditions, which are characterized by high UF_6 speed ratios and a small difference between the ternary and the binary diffusion coefficients, considerably exceed the upper bound of the equilibrium separation effect (cf. Sect.4.1.2 and Fig. 4.5).

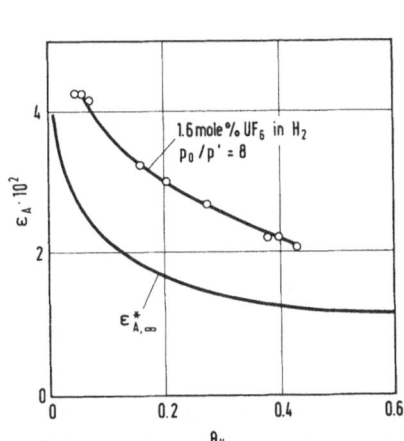

Fig.4.10. Exceeding the upper bound of the equilibrium separation effect $\varepsilon^*_{A,\infty}$ in the standard separation nozzle

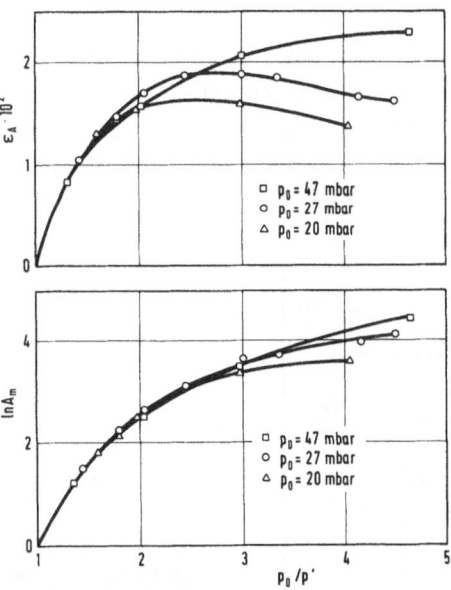

Fig.4.11. Elementary effect ε_A of isotope separation and separation factor A_m of H_2/UF_6 mixture versus expansion ratio p_0/p' in the standard separation nozzle at various inlet pressures p_0 ($N^0_u = 0.03$, $\theta_u = 0.25$, $p' = p''$)

Figure 4.11 shows the elementary effect of isotope separation and the separation factor of the mixture at a constant UF_6 cut for a mixture of H_2/UF_6 with 3 mole % of UF_6, plotted versus the expansion ratio for various inlet pressures, p_0. At the low inlet pressures, the elementary effect of isotope separation passes through a maximum when the expansion ratio rises; the separation factor of the mixture, however, rises continuously with the expansion ratio. This behavior can be explained by the fact that the maximum of isotope separation tends to be shifted towards smaller angles of deflection with increasing expansion ratio, because the increasing speed ratio causes isotope separation to proceed more quickly over the angle of deflection. Consequently, at high expansion ratios and low inlet pressures, the maximum of isotope separation is passed upstream from the skimmer edge. By increasing the inlet pressure p_0, the process of isotope separation can be slowed down in accordance with the decreasing Knudsen number; the maximum of isotope separation is then shifted downstream. For high inlet pressures, maximum isotope separation cannot be reached upstream from the skimmer and there is a monotonic increase of ε_A with p_0/p' (Sect.4.2.3).

Direct measurement of the development of isotope separation in the flow field of the separation nozzle is possible by means of free molecular probes /55/. This measuring technique, which is described in greater detail in Sect.5.3.1, allows the local values of the fluxes of the components of the mixture to be determined and the partial cuts to be obtained from these values by integration. Moreover, one obtains the local values of the speed ratios of the components of the mixture, from which the equilibrium separation effect can be calculated.

Figure 4.12 shows the results of measurements performed by molecular probes in the standard separation nozzle using SF_6 instead of the corrosive UF_6 with He as a

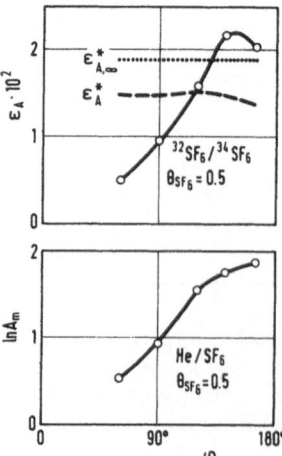

Fig.4.12. Azimuthal development of the elementary effect ε_A of isotope separation between $^{32}SF_6$ and $^{34}SF_6$, and of the separation factor A_m between He and SF_6 for the SF_6 stream surface $\theta_{SF_6} = 0.5$. (Results of free molecular probe measurements performed on the standard separation nozzle operated with a He/SF_6 mixture with 1 mole % SF_6 at an expansion ratio of $p_0/p' = 3.1$)

light auxiliary gas. It is seen that the elementary effect ε_A of isotope separation passes through a maximum over the angle of deflection, while the gas separation factor A_m between He and SF_6 rises continuously. The values measured for ε_A not only are clearly above the values of equilibrium separation following from the SF_6 speed ratios but, in the flow region upstream the skimmer, also exceed the upper limit of equilibrium separation.

5. Analysis of Flow Processes in the Separation Nozzle

5.1 General Characteristics and Flow Parameters

The relative change in average isotopic ratios in the partial streams on both sides of a UF_6 stream surface depends on the transport of isotopes through the stream surface and on the azimuthal flow rates of the partial streams, i.e., the respective isotope transport in the flow direction. The ratio of the separation transport streams perpendicular to the flow direction and the convective isotope transport in the flow direction consequently determines the elementary effect of isotope separation. Analogously, the separation factor of the mixture is determined by the ratio of the opposed transport streams of UF_6 and auxiliary gas normal to a molar stream surface of the mixture to the azimuthal transport streams of the mixture on both sides of this stream surface.

The ratio of transport streams which determines the spatial separation of isotopes and mixture components converges against zero, if one looks at the limit cases of continuum flow and free molecular flow. The reason is that in case of a continuum flow the radial transport stream based on diffusion is negligibly small relative to the azimuthal transport, unless the angle of deflection of the flow is made very large. In a steady-state free molecular flow, the spatial distribution of molecules is independent of the molecular mass if, as in the separation nozzle, no external forces act on the molecules and there is no interaction of the molecules with a moving wall. Consequently, finite values for isotope and mixture separation will be found only in the transition regime between continuum flow and free molecular flow.

To describe the properties of the separation nozzle flow, it is advisable to use various characteristic numbers, such as the Knudsen number Kn, the Reynolds number Re, the Mach number Ma, and the speed ratio S.

In the Knudsen number,

$$Kn = \lambda/a \tag{5.1}$$

(λ = mean free path), the characteristic dimension used as a basis below is the nozzle width, $a = r_0 - r_1$, i.e., the radial distance between the outer and inner boundaries of the confined flow. The Reynolds number,

$$Re = \bar{\rho} \, \bar{v} \, a/\eta \tag{5.2}$$

($\bar{\rho}$ = mean mass density, $\bar{v}$ = mean flow velocity, η = dynamic viscosity), is identical in slit-type separation nozzle systems with the ratio of the mass flow per unit length of the nozzle slit and the dynamic viscosity. The Knudsen number, the Reynolds number and the mean Mach number of the flow can approximately be linked by the relationship

$$Kn = \sqrt{\gamma} \, Ma/Re \tag{5.3}$$

if problems are neglected which, e.g., result from local changes in the Mach number in the flow field or in defining properly the mean free path in the gaseous mixture.[5]

The Knudsen number Kn^+ optimal for isotope separation is defined, as is shown in Sect.4.2.3, by the transient maximum of the elementary effect of isotope separation being reached right at the splitting point of the flow by the skimmer. On the basis of the angle of deflection of 180°, which is typical of most separation nozzle systems, and the usual operating conditions of the separation nozzle process (Sect.2), a characteristic Knudsen number of approximately 10^{-2} is found for the separation nozzle flow. Under those conditions, the Reynolds number of the separation nozzle flow is around 100, i.e., the flow is laminar and highly viscous.

If, for simplification, it is assumed that the separating pressure diffusion stream increases with the square of the UF_6 speed ratio S_u, and hence of the average Mach number Ma_m of the mixture (Sect.4.2), and that the convective azimuthal transport through the nozzle increases linearly with the Reynolds number, then the influence of the Reynolds number on isotope and mixture separation can be described qualitatively by means of the parameter $\overline{Ma}_m^2/Re$. At high Reynolds numbers, when viscous effects can largely be neglected, the Mach number of the mixture is practically independent of the Reynolds number; at a given angle of deflection,

---

[5] For better distinction, the Knudsen number will mainly be used below to characterize such effects as are directly related to the molecular structure of the gas mixture, e.g., to the molecular velocity distribution or the development of the diffusion processes. The Reynolds number is mainly employed to characterize viscous effects and pressure losses in the separation nozzle flow.

the separation of the isotopes and of the mixture components decreases inversely proportional to Re. At low Reynolds numbers, viscous effects become more and more important. The Mach number of the mixture decreases rapidly with decreasing Re, and the ratio of transport streams characterized by $\overline{Ma}_m^2/Re$ becomes smaller. If conditions optimal for separation are characterized by the maximum value of $\overline{Ma}_m^2/Re$, it is evident that the corresponding Mach number must be considerably below the value that would be attained in an isentropic flow at the same expansion ratio.

If the separation nozzle system shown in Fig.2.1 is operated, e.g., at an expansion ratio of 2 with an H_2/UF_6 mixture with 3 mole % of UF_6 at a Reynolds number of 100, which would roughly correspond to the optimum economic operating conditions of the system, the average Mach number of the mixture of $\overline{Ma}_m = 0.4$ is approximately half the value attainable in the case of isentropic expansion. A Mach number of the mixture around 1 will be attained only in a small subregion of the flow field. Consequently, under the operating conditions characteristic of the process, the separation nozzle flow is subsonic and in some subregions, transsonic /33,50/.

Because of the low mean molecular weight of the mixture, the flow velocity is many times higher than the velocity of sound in pure UF_6, but it is small compared with the velocity of sound in the light auxiliary gas. This condition can be illustrated by a supersonic UF_6 jet of low number density being embedded in a subsonic jet of high number density formed by the auxiliary gas. Since the resultant jet of the mixture has the properties of a subsonic flow, the UF_6 can be deflected by any angle without there being any deceleration or generation of entropy due to shock waves. Even at high expansion ratios, at which supersonic regions occur in the flow field, shock waves are largely suppressed; this is due to the relatively high Knudsen number, where the shock thickness would become comparable with the characteristic dimensions of the flow and, consequently, a discontinuity like a shock cannot fully develop.

5.2 Calculating the Flow Field for Uniform Gases

Since the separation nozzle flow belongs to the transition regime between continuum and free molecular flow conditions and the strong separation between UF_6 and the light auxiliary gas greatly affects its behavior, calculating the flow field is very difficult. For this reason, the theoretical investigations performed so far were limited to solving the Navier-Stokes equations for a uniform gas. For simplification, a curved slender channel, i.e., a nozzle contour without any abrupt changes, was used as a basis instead of the actual nozzle geometry (Fig.2.1) /56,57/. Since, in this case, changes in the flow properties in the azimuthal direction are small compared with those in the radial direction, higher-order azimuthal terms were neglected in the solution of the Navier-Stokes equation.[6]

In calculating the flow field in the separation nozzle, it was assumed on the basis of experimental results that a flow with a trapezoidal velocity profile exists at the beginning of deflection ($\phi = 0$). At the Reynolds number of the nozzle flow optimal for isotope separation ($Re \cong 100$), this initial profile changes into an approximately parabolic velocity profile typical of a fully developed viscous flow within an angle of deflection of approximately 30°. At higher Reynolds numbers ($Re \cong 400$), a velocity profile is developed in the core of the flow unaffected by boundary layers, which is similar to that of a potential vortex, i.e., the velocity changes with the reciprocal radius. The average flow velocity in the range of Reynolds numbers considered here clearly increases with increasing Reynolds number; at $Re = 400$ it is approximately 30% higher than at $Re = 100$. Typical velocity profiles resulting after deflection of the flow by an angle of 60° are shown in Fig.5.1 for $Re = 100$ and $Re = 400$.

The temperature profile calculated by the simplified model equations roughly corresponds to that of an isoenergetic flow, where the static temperature T and the Mach number Ma are interrelated by

$$T = T_0 / (1 + \frac{\gamma-1}{2} Ma^2) \quad , \tag{5.4}$$

(T_0 = reservoir temperature = temperature of the gas before expansion, γ = ratio of specific heats). A more detailed analysis of the temperature development shows that

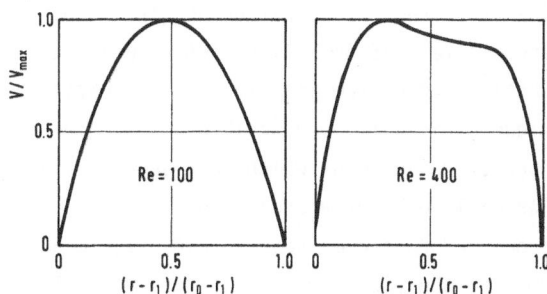

Fig.5.1. Radial velocity profiles for a uniform gas in a curved slender channel at various Reynolds numbers (angle of deflection $\phi = 60°$, expansion ratio $p_0/p' = 4$)

---

[6] However, in this simplification the system of elliptic differential equations of the complete Navier-Stokes equations changes into a system with parabolic properties. For this reason, it is not possible to postulate the boundary conditions in the suction channels of the separation nozzle; they are determined by the conditions at the nozzle inlet and at the nozzle walls.

the local stagnation temperature of the gas in the outer regions of the curved flow is slightly above the reservoir temperature T_0, and in the inner regions, slightly below T_0. This is due to the fact that the inner regions of the curved flow perform work on the outer flow regions by viscous sweeping.

At the low Reynolds numbers typical of the separation nozzle method, the average Mach number of the flow is clearly lower, because of dissipative losses, than the value attainable in the case of isentropic expansion. These losses, which are best characterized by the increase in entropy of the flowing gas, have been determined by means of the dissipation function from the velocity and temperature fields calculated by means of the simplified Navier-Stokes equations. The increase in entropy of a pure gas which is expanded by a factor of 4 at Re = 100 in a curved channel similar to the standard separation nozzle is roughly half that of the maximum increase in entropy

$$\Delta s_{max} = R \ln(p_0/p') \quad .$$
(5.5)

This maximum value corresponds to isenthalpic throttling of the gas from the inlet pressure p_0 to the suction pressure p' of the light fraction. At Re = 100, the increase in entropy due to viscous effects is about four times higher than the increase due to thermal conduction in the flow field /56,57/.

5.3 Experimental Studies of the Separation Nozzle Flow

5.3.1 Methods of Measurement

For experimental investigation of the flow processes in separation nozzles, free molecular probes have been given preference so far /45,50/. Since interpretation of the results obtained in this way is facilitated by familiarity with the basic principle of the probe measurement technique, this will be described in more detail below. In addition, some other methods of measurement which have been applied in model experiments on separation nozzles will be briefly explained.

The principle of the free molecular probe method can be seen in Fig.5.2 /58,59/. The probe consists of a tube aligned normal to the flow, whose diameter is small relative to the mean free path in the flow so that the behavior of the flow is practically not affected by the probe. The tube sealed at the top has an opening in the side, which can be oriented at any angle relative to the flow direction by turning the probe around the tube axis. The gas entering the opening is fed into a mass spectrometer, and the ion currents I_i for the various components i of the

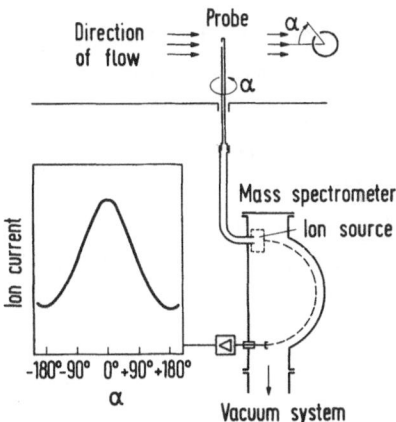

Direction of flow

Probe

Ion current

-180°-90° 0°+90°+180°

α

Mass spectrometer
Ion source

Vacuum system

Fig.5.2. Principle of free molecular probe technique with selective detection of the components of a gaseous mixture by means of a mass spectrometer

gas mixture are recorded as a function of the angle of attack α of the probe open-ing. The angular dependence of the ion current is shown schematically in Fig.5.2.

Under the simplified assumption that the opening of the probe corresponds to an ideal orifice, i.e., a hole in an infinitesimally thin wall, and that the flow is in a state of local thermodynamic equilibrium, the following expression is found for the flow rate of the component i into the opening of the probe /60/

$$I_i' = \frac{F \nu_i c_i}{2\sqrt{\pi}} \{\exp(-S_i^2\cos^2\alpha_i) + \sqrt{\pi}S_i [1+\mathrm{erf}(S_i\cos\alpha_i)]\cos\alpha_i\} \tag{5.6}$$

with

$$\mathrm{erf}(x) = \frac{2}{\sqrt{\pi}} \int_0^x e^{-y^2} dy \quad,$$

where ν_i is the number density, c_i the most probable thermal velocity, S_i the speed ratio, α_i the angle of attack of the probe opening relative to the direction of flow of the component i of the gas mixture, and F is the area of the probe opening. If the number density in the mass spectrometer is sufficiently low, the ion current I_i is proportional to the flow rate I_i' into the opening of the probe. It follows direct-ly from (5.6) that measurement of the ion current at various angles of attack α_i al-lows the speed ratio S_i of the component i of the mixture and its local flow direc-tion to be determined. Usually, the speed ratio is evaluated from the ion current ratio

$$I_i(\alpha_i=0)/I_i(\alpha_i=90°) = \exp(-S_i^2)+\sqrt{\pi} S_i[1+\mathrm{erf}(S_i)] \quad. \tag{5.7}$$

Evidently, this corresponds to the determination of the local Mach number from the local values of stagnation pressure and static pressure under continuum flow condi-tions. The relative difference in ion currents at $\alpha_i = 0$ and $\alpha_i = 90°$ is a measure of the ratio of the flow velocity and the thermal velocity of the molecules. If the local velocity distribution is strongly perturbated, the S-values formally determined from (5.7) can therefore be considered as effective values of the speed ratio /58/ (cf. also Sect.5.3.3).

The local flux j_i of a component of the mixture is derived from the net stream through an area located normal to the flow direction. This net stream is equivalent to the difference in the streams passing through the opening of the probe at the angles of attack $\alpha_i = 0$ and $\alpha_i = 180°$, if that opening represents an ideal orifice. Correspondingly, it holds that

$$j_i = \text{const} \{I_i(\alpha_i=0)-I_i(\alpha_i=180°)\} \quad . \tag{5.8}$$

The constant in (5.8) must be determined by a calibration measurement in a stagnant gas of known density and temperature. The flux profiles of the various components of the mixture in a flow cross section can be used to determine, by integration in accordance with (4.6), the partial cuts θ_i and the spatial development of the molar stream surfaces. From these quantities, the separation factor and the separation effect can be determined so that the spatial development of separation can be evaluated besides the development of the speed ratios in the separation nozzle.

From the relation

$$I_i'(\alpha_i=90°) = \frac{F \, \nu_i c_i}{2\sqrt{\pi}} \quad , \tag{5.9}$$

it follows with

$$p_i = \nu_i kT$$

and

$$c_i = \sqrt{2kT/M_i}$$

that, for a given static temperature, measurement of the ion current

$$I_i(\alpha_i=90°) = \text{const} \, p_i/\sqrt{T} \tag{5.10}$$

in a calibrated probe measuring setup allows the partial pressure p_i to be determined. The local static temperature T of the flowing gas mixture can be calculated directly from the local speed ratios and fluxes of the components and the reservoir temperature T_0, if the flow can be assumed to be isoenergetic (5.4).

If the components of the flowing gaseous mixture are assumed to be in a local thermodynamic equilibrium, i.e., to have equal flow velocities and equal tempera-

tures, the local properties of the mixture can be calculated simply from the characteristics of the components. In determining the local molar fractions N_i, the speed
ratio of the mixture S_m and the static pressure p of a mixture with x components,
the following relations are normally applied:

$$N_i = j_i / \sum_1^x j_k \quad , \tag{5.11}$$

$$S_m = S_i \sqrt{\overline{M}/M_i} \quad , \tag{5.12}$$

with

$$\overline{M} = \sum_1^x N_k M_k \quad ,$$

$$p = \sum_1^x p_k \quad . \tag{5.13}$$

In flow regions characterized by large changes of state within a few mean free paths,
the components of the mixture may differ considerably with respect to their flow velocities and kinetic temperatures, i.e., their mean molecular energies of thermal
motion. This is particularly true of highly different molecular masses of the components of a mixture because, as a result of the ineffective exchange of energy,
long relaxation times will result for the adjustment of thermodynamic equilibrium
among the components. In this case, evaluating the measurements in accordance with
(5.11-13) can assess the behavior of the mixture only in qualitative terms.

Normally, the molecular probes used to investigate separation nozzle flows have
diameters of 0.4 mm, which ensures both satisfactory mechanical stability and sufficiently short response times for the measuring setup. The diameter of the lateral
opening is about 0.02 mm. For reasons of manufacturing, it is not possible at these
small dimensions to achieve a shape of the opening resembling an ideal orifice with
infinitesimally small wall thickness. The lateral opening in the wall of the probe
has the shape of a short cylindrical channel, whose molecular flow resistance must
also be taken into account when evaluating probe measurements /58,61/.

The radius of deflection of the separation elements used for free molecular probe
measurements is 100 to 200 mm, which makes it more than three orders of magnitude
larger than that of commercial separation nozzle elements. Because of these large
characteristic dimensions, the operating pressure optimal for separation is some
0.1 mbar, ensuring an approximately molecular flow with regard to the probe. Since
the large characteristic dimensions require extremely high volume flows to be extracted from the separation element and since the pumping systems available cannot
be run on the radioactive and highly corrosive UF_6, the probe measurements use chemically stable model gases instead of UF_6, such as SF_6, whose structure is similar to

that of UF_6, or a compound with the same molecular weight, C_7F_{14} (perfluoromonomethyl cyclohexane).

Besides the molecular probe with a mass spectrometer, a number of other methods of measurement have been used to study separation nozzle flows; they allow additional information to be obtained or certain problems to be clarified with less expenditure in terms of measuring equipment or time. These additional measuring methods will be explained briefly below.

For direct measurement of the local static temperature of the separation nozzle flow, free molecular temperature probes have been used; they normally consist of a thermally insulated cylinder aligned normal to the flow /45/. The equilibrium temperature of a cylinder in a free molecular flow depends on the local static temperature T of the flowing gaseous mixture, and the speed ratios S_i, the fluxes j_i, and the thermal accommodation coefficients of the components of the mixture, so that T can be reduced to quantities which are directly accessible experimentally.

For selective measurements of the local number density and the temperature of the heavy component of the mixture, the absorption of laser radiation which is passed through the gaseous mixture normal to the direction of flow is utilized /62,63/. A tunable CO_2 laser is applied, in the spectral range of which the SF_6 and C_7F_{14} compounds used as model gases instead of UF_6 show strong absorption bands. Since the temperature dependence of the absorption coefficient changes greatly if the laser radiation is tuned to different regions of the absorption band, absorption measurements with several discrete laser lines allow the local density and the local rotational-vibrational temperature of the absorbing component of the mixture to be determined. One particular advantage of absorption measurements lies in its short response time so that fast transient processes, such as flow instabilities in separation nozzle systems with gas dynamic flow deflection (Sect.8.3.4), can be resolved.

Information about dissipative losses in the separation nozzle flow can be derived with relatively little experimental expenditure by simple pressure measurements /57,64/. For this purpose, the stagnation and static pressure profiles are determined in a flow cross section by means of a Pitot probe and by means of pressure taps in the sidewalls confining the slit-shaped separation element. Because of the low Reynolds number of the separation nozzle flow, the so-called fishmouth-type probes with a flattened opening extended in the direction of the nozzle slit must be used instead of conventional Pitot probes with circular openings.[7] Under the simpli-

---

[7] In a conventional Pitot probe, the pressure measured in a subsonic flow differs greatly from the local stagnation pressure, if the ratio Re/Ma of the Reynolds number formed with the width of the probe opening and the Mach number of the flow falls below a value of 10. In a fishmouth-type probe, the values of this ratio may be three times smaller /64/.

fied assumption of an isoenergetic flow corresponding to (5.4) prevailing in the
separation nozzle, the increase in entropy, which characterizes the dissipative
losses, can be determined from the total pressure loss weighted by the flux j. The
increase in entropy determined in this way is normally referred to the increase in
entropy Δs_{max} under conditions of isenthalpic throttling from the nozzle inlet pressure p_0, to the suction pressure p' of the light fraction. As a measure of the dissipative losses in terms of the relative increase in entropy one then obtains

$$\Delta s_{rel} = \frac{1}{\ln(p_0/p')} \int_{r_1}^{r_0} j \ln(p_0/p_s)dr / \int_{r_1}^{r_0} j \, dr \qquad (5.14)$$

with p_s representing the local stagnation pressure of the flowing gas mixture. The
flux j is calculated from the readings of the local static pressure and the stagnation pressure, taking into account the local composition of the mixture in accordance with well-known gas dynamics relations. For the sake of completeness it
should be added that when determining the increase in entropy from the average pressure loss, the reduction in entropy to be assigned to the spatial separation of the
components of the mixture is not taken into account. Since, in addition, an isoenergetic flow is assumed in (5.14), deviations of the local stagnation temperature from
the reservoir temperature T_0 of the gas are also neglected. However, these neglected
aspects are only of minor importance for the analysis of dissipative losses, because
both terms are small compared with the increase in entropy brought about by viscous
effects, as can be shown by free molecular probe measurements or by simple estimates.

5.3.2 Flow Field and Spatial Development of Separation

In order to illustrate the behavior of the flow and the spatial development of separation in the curved nozzle, some typical results of free molecular probe measurements will be described below. The separation nozzle used in the probe measurements,
whose contour is shown in Fig.5.3 and which has a radius of deflection of $r_0 = 150$ mm,
was run on a He/C_7F_{14} mixture with 4 mole % of C_7F_{14} at equal expansion ratios
$p_0/p' = p_0/p'' = 2.5$ for the light and heavy fractions. The inlet pressure p_0 was
0.12 mbar, which corresponds to the value optimal for separation of the mixture
under the conditions listed above. The Reynolds number of the flow was around 50.

In Figs.5.4-7 the radial profiles of the partial speed ratios S_a, S_h and the
fluxes j_a, j_h of the auxiliary gas and the heavy component of the mixture, the
speed ratio of the mixture S_m, and the static pressure p_m, of the mixture are shown
for three flow cross sections at the angles of deflection $\phi = 0°$, 90°, and 160°.
Moreover, the separation factor A_m between He and C_7F_{14} is plotted as a function
of the partial cut θ_h of the heavy component of the mixture.

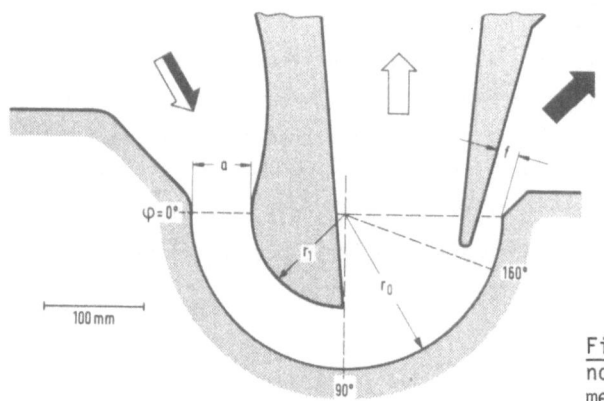

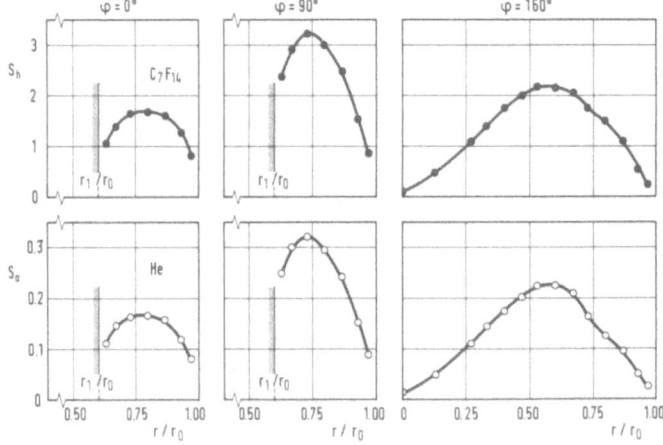

Fig.5.3. Contour of the separation nozzle used for free molecular probe measurements

Fig.5.4. Results of free molecular probe measurements on the radial profiles of the partial speed ratios S_a and S_h of the light auxiliary gas and the heavy component of the gaseous mixture at various angles of deflection ϕ of a standard separation nozzle (96 mole % of He/4 mole % of C_7F_{14}, $p_0/p' = p_0/p'' = 2.5$, Re $\cong$ 50; for the nozzle contour, see Fig.5.3)

Figure 5.4 shows that the radial profiles of the partial speed ratios S_a and S_h of the light auxiliary gas and the heavy component of the mixture are largely similar. In relative terms, S_a and S_h differ by a factor roughly corresponding to the square root of the ratio of molecular weights; accordingly, roughly equal flow velocities and equal thermal energies of translational motion can be assumed for the two components of the mixture.[8] It also follows from the probe measurements that

---

[8] This does not mean that the flowing gas mixture is in a state of local thermodynamic equilibrium. In regions with a strong gradient of the speed ratio, the local molecular velocity distribution differs greatly from a Maxwellian distribution, as will be shown in Sect.5.3.3.

the flow is accelerated in the first half of the deflection, where it is confined by the inner wall of the nozzle and the deflection wall. The maximum speed ratio S_h of the heavy component of the mixture rises from 1.7 at $\phi = 0°$ to 3.2 at $\phi = 90°$ under the experimental conditions prevailing in this case. The radial profile of the speed ratio has a relatively broad maximum at the inlet, which is typical of viscous intake flows. At $\phi = 90°$, the characteristic parabolic profile of a fully developed viscous flow is observed. In the second half of deflection, in which the flow is confined by the deflection wall only, the width of the gas jet increases strongly and the maximum speed ratio of the heavy component decreases to a value of $S_h \cong 2.2$ in the flow cross section upstream from the skimmer ($\phi = 160°$).

As a consequence of the relatively high Knudsen number of the separation nozzle flow and the partly very high radial velocity gradients, there is a considerable velocity slip along the solid walls of the nozzle. Depending upon the radial gradients of S_h and S_a at the wall, the speed ratio encountered right at the wall is between 2% and 10% of the maximum speed ratio in the respective flow cross section, as has been determined by measurements with a molecular probe integrated in the wall /45/. It should be added that the large jump in the speed ratio at the inner wall of the nozzle at $\phi = 90°$ must not be directly interpreted as an extremely high velocity slip; it is rather due to the sudden expansion of the gas at the edge of the inner nozzle wall, which terminates at approximately $\phi = 90°$ (Fig.5.3).

The profiles of the partial fluxes j_a and j_h shown in Fig.5.5 directly indicate the development of the spatial separation between the auxiliary gas and the heavy component of the mixture. At the beginning of deflection ($\phi = 0°$), the flux profiles normalized to their respective maximum values differ only very little, their shapes roughly corresponding to the profiles of the partial speed ratios S_a and S_h; con-

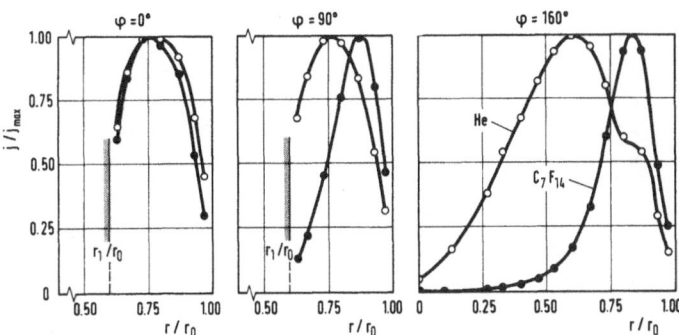

Fig.5.5. Results of free molecular probe measurements on the radial profiles of the partial fluxes j_a and j_h of the light auxiliary gas and the heavy component of the gaseous mixture at various angles of deflection ϕ of a standard separation nozzle (96 mole % of He/4 mole % of C_7F_{14}, $p_0/p' = p_0/p'' = 2.5$, Re $\cong$ 50; for the nozzle contour, see Fig.5.3)

sequently, the profiles of j_a and j_h and of S_a and S_h mainly reflect the flow velocity. Since the curvature of the streamlines in the inlet section of the separation nozzle is opposite to the curvature of the deflection wall, the light auxiliary gas is slightly enriched near the deflection wall at $\phi = 0°$; accordingly, the
normalized flux of the light auxiliary gas is slightly higher in the flow region
at the deflection wall than that of the heavy component. For larger angles of deflection, the flux profiles of the components of the mixture differ increasingly.
The radial profile of j_a continues to correspond roughly to that of S_a and the flow
velocity in accordance with the low speed ratio of the auxiliary gas and the resultant slight radial change in partial density. However, the profile of j_h differs
considerably from that of S_h because of the strong radial increase in number density of the heavy component of the mixture. In the second half of the deflection, the
heavy component is concentrated in the flow region at the deflection wall, in which
the speed ratio S_h drops in the radial direction. The influence exerted on the flow
by the skimmer can be recognized particularly in the flux profile of the light component at $\phi = 160°$, which has a step shape in the region upstream the skimmer ($r/r_{\ddot{0}} =$
0.8).

The radial profiles of the speed ratio S_m and the static pressure p of the He/C_7F_{14}
mixture are shown in Fig.5.6 for various flow cross sections. At the beginning of
deflection ($\phi = 0°$), the profile of S_m is similar to those of S_a and S_h; the maximum
value of S_m is about 0.4. The static pressure p_m of the mixture is roughly constant
in the radial direction at $\phi = 0°$ and is about 25% below the nozzle inlet pressure

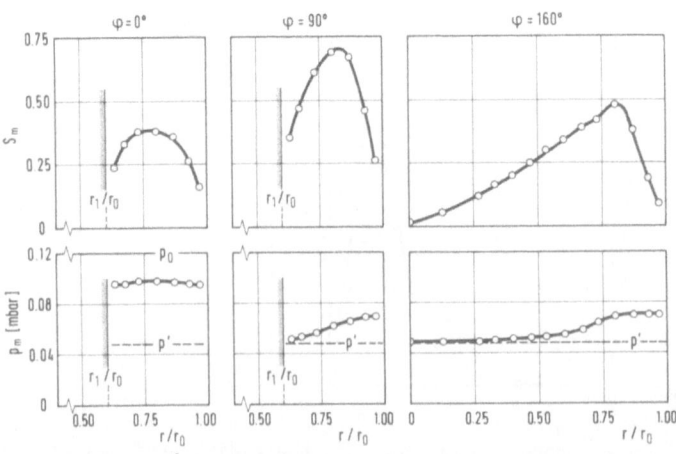

Fig.5.6. Results of free molecular probe measurements on the radial profiles of the
speed ratio S_m and the static pressure p_m of the gaseous mixture at various angles
of deflection ϕ of a standard separation nozzle (96 mole % of He/4 mole % of C_7F_{14},
$p_0/p' = p_0/p'' = 2.5$, Re ≈ 50; for the nozzle contour, see Fig.5.3)

p_0. Unlike the behavior of an isentropic flow, in which a change in the Mach number and, hence, in the speed ratio is possible only by a change in the flow cross section, the speed ratio in the highly viscous separation nozzle flow rises strongly in the nozzle channel formed by the inner nozzle wall and the deflection wall, which channel has an azimuthally constant cross section according to Fig.5.3. The increase in the speed ratio in this case correlates with the pressure loss due to dissipation, i.e., the decrease of density due to viscous effects is offset by a corresponding increase in flow velocity in the steady-state flow. At $\phi = 90°$ a maximum speed ratio of the mixture of about 0.7 is reached; the static pressure at the inner boundary of the flow ($r = r_1$) has dropped approximately to the suction pressure p' of the light fraction and increases in the radial direction in the way typical of a curved flow.

It also follows from Fig.5.6 that the radial difference Δp between the static pressures at the deflection wall and at the inner boundary of the jet ($r = r_1$ or $r = 0$) remains roughly constant in the second half of deflection. Consequently, the azimuthal decrease of the flow velocity v is compensated by the broadening of the flow cross section so that there is almost no change in the intensity of the separating centrifugal field between $\phi = 90°$ and $\phi = 160°$, as characterized by the differential pressure,

$$\Delta p = \int_0^{r_0} \rho \, \frac{v^2}{r} \, dr \quad . \tag{5.15}$$

The profiles of the speed ratio of the mixture deviate more and more strongly from those of the partial speed ratios as deflection increases. The maximum of S_m lies at greater radii than those of S_a and S_h, because the molar fraction of the heavy component flowing at a high speed ratio increases towards the deflection wall.

The spatial development of the separation of the mixture can be seen from Fig.5.7, where the logarithm of the separation factor of the mixture, $\ln A_m$, is plotted versus the partial cut θ_h of the heavy component. At the beginning of deflection ($\phi = 0°$), the heavy component is enriched in the flow region near to the inner nozzle wall and $\ln A_m$ is negative, since the curvature of the streamlines in the nozzle inlet is reverse to that of the deflection wall. As deflection proceeds, a correlation

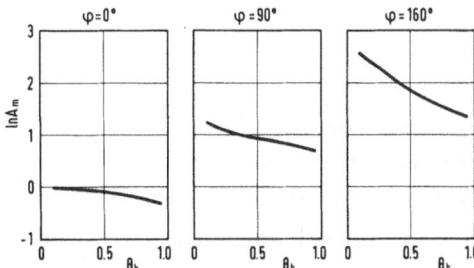

Fig.5.7. Results of free molecular probe measurements of the correlation between the separation factor A_m of the mixture and the cut θ_h of the heavy component at various angles of deflection ϕ of a standard separation nozzle (96 mole % of He/4 mole % of C_7F_{14}, $p_0/p' = p_0/p'' = 2.5$, Re $\cong$ 50; for the nozzle contour, see Fig.5.3)

48

between the separation factor and the cut develops, which is typical of separation
in a centrifugal field and is characterized by a continuous rise of A with decreasing θ.

If SF_6 is used instead of C_7F_{14} as a model gas for UF_6, it is possible by means
of the molecular probe method to measure not only the separation between the light
auxiliary gas and the heavy model gas, but also isotope separation, i.e., separation
between the $^{32}SF_6$ and $^{34}SF_6$ molecules in the flow field of the separation nozzle.
Some typical results of such measurements are shown in Fig.5.8. This is a plot

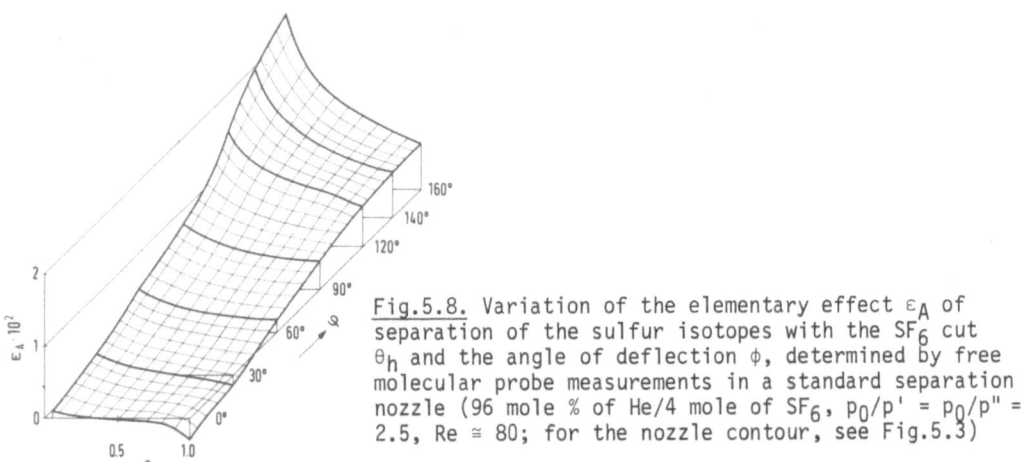

Fig.5.8. Variation of the elementary effect ε_A of
separation of the sulfur isotopes with the SF_6 cut
θ_h and the angle of deflection ϕ, determined by free
molecular probe measurements in a standard separation
nozzle (96 mole % of He/4 mole of SF_6, $p_0/p' = p_0/p'' =$
2.5, Re ≅ 80; for the nozzle contour, see Fig.5.3)

of the elementary effect ε_A of separation of the sulfur isotopes as a function of
the SF_6 cut θ_h and the angle of deflection ϕ. As in the separation between He and
C_7F_{14} (Fig.5.7), a reverse separation is observed at the beginning of deflection;
the elementary effect ε_A of isotope separation is negative for $\theta_h > 0.5$. This reverse separation vanishes in the flow regions at the deflection wall at an angle
of deflection ϕ of approximately 45°. In the further course of deflection, ε_A at
the deflection wall ($\theta_h = 1$) rises continuously up to an angle of deflection of approximately 120°, while no major increase can be observed between $\phi = 120°$ and $\phi = 160°$.
In the region of low SF_6 cuts, the increase in ε_A with the angle of deflection is
much more pronounced than in the case of high cuts. Near the end of deflection, the
dependence of the separation effect ε_A on the cut θ_h is typical of isotope separation in a centrifugal field and is roughly similar to the relation pertaining to
equilibrium separation in accordance with (4.12) (cf. also Fig.4.5,10).

Some typical aspects concerning the temperature development in the separation
nozzle flow are shown in Fig.5.9 /45/. The measurements were performed by means of
free molecular temperature probes. The gases were pure helium and a mixture of He
and SF_6 with 7.6 mole % of SF_6, the expansion ratio was 4, the Reynolds number of
the flow was 80. The ratio of the nozzle width a to the radius of curvature r_0 of

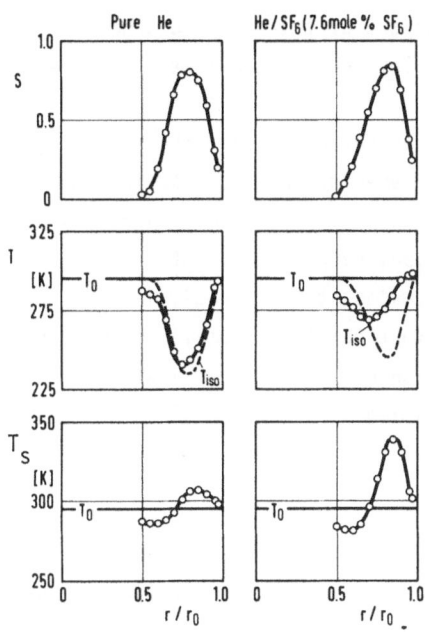

Fig.5.9. Comparison of the radial profiles of the static temperature T and the stagnation temperature T$_s$ for pure He and a He/SF$_6$ mixture in a standard separation nozzle at the angle of deflection $\phi = 90°$ at approximately identical speed ratios for the pure gas and the mixture (p$_0$/p' = p$_0$/p" = 4, Re $\cong$ 80)

the deflection wall in the separation nozzle used here was a/r$_0$ = 0.3 mm, which is 25% smaller than in the arrangement shown in Fig.5.3.

The diagrams in Fig.5.9 are plots of the radial profiles of the speed ratio, the local static temperature and the local stagnation temperature for pure He and for the He/SF$_6$ mixture at the angle of deflection $\phi = 90°$. The dashed curves reflect the temperature calculated from the speed ratio for an isoenergetic change of state in accordance with (5.4). In the pure gas, the static temperature T of the flow in the region close to the deflection wall is slightly above the isoenergetic temperature T$_{iso}$, while in the inner regions of the jet, T remains under T$_{iso}$. Accordingly, also the local stagnation temperature T$_s$ is higher in the flow region close to the deflection wall and lower than the reservoir temperature T$_0$ in the inner flow region. As mentioned in Sect.5.2, this temperature development results from the fact that the inner regions of the curved flow perform work on the outer flow regions by viscous sweeping.

In the mixture, unlike the uniform gas, the radial profile of the static temperature differs considerably from the isoenergetic temperature profile. In the flow region at the deflection wall, T is clearly higher than T$_{iso}$; the stagnation temperature T$_s$ correspondingly exceeds the reservoir temperature T$_0$. Right at the deflection wall, a temperature jump is observed (T $\neq$ T$_0$) between the flow and the wall, in which the temperature of the mixture exceeds the wall temperature by some 5 K. The increase in the temperature of the mixture in the flow region at the deflection wall is clearly more pronounced than the corresponding temperature increase of the uniform gas. This can be explained by the radial transport of the heavy component, which has increased in its partial stagnation enthalpy due to acceleration by the light auxiliary gas.

With respect to the diffusion processes in the separation nozzle, the following conclusions can be drawn from the temperature measurements. The increase in temperature in the outer flow regions and the decrease of temperature in the inner flow regions is associated with an increase and a decrease, respectively, of the diffusion coefficient, which accelerates separation processes at large radii and delays them for small radii. Moreover, it must be stated that relatively small temperature gradients occur in the mixture even at high expansion ratios, which means that thermal diffusion has only a slight influence on separation.

5.3.3 Non-Equilibrium Effects

In Sect.4.2 it was pointed out that the molecular velocity distribution is strongly perturbated over large regions of the separation nozzle flow and cannot even approximately be described by a Maxwellian distribution. The essential reason for these perturbations is seen in the fact that because of the relatively high Knudsen number of the flow and the high velocity gradients in the flow field, there are large changes of state within a mean free path. In this case, a volume element of the flow, in which the center-of-mass velocities, partial densities and mean values of the thermal energies of the components of the mixture differ only slightly from the corresponding values in adjacent volume elements, must be extremely small. As a result, the number of molecular collisions within the volume element is very low and no local thermodynamic equilibrium can be established.

Furthermore, it must be taken into account that the exchange of translational energy between the UF_6 molecules and the molecules of the auxiliary gas is much weaker, because of the high mass ratio, than the exchange of energy between molecules with identical masses. Consequently, local thermodynamic equilibrium can be established much more quickly within each individual species than between the components of the mixture. Even if the two types of molecules have approximately Maxwellian distributions, marked local differences may yet occur in temperatures and in the center-of-mass velocities of the components of the mixture /43/.

The phenomena originating from the mechanisms mentioned above, which are normally called "non-equilibrium" or "rarefaction" effects, can hardly be described theoretically for separation nozzle flows because, at present, there are no suitable methods for solving the Boltzmann equation for a strongly perturbated velocity distribution in disparate mass mixtures. Merely Monte-Carlo calculations have been used so far to derive some qualitative information about the different behavior of the

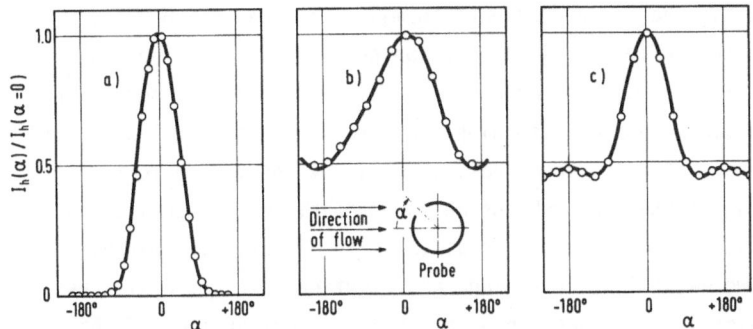

Fig.5.10. Typical results of free molecular probe measurements of the dependence of the normalized ion current of the heavy component ($I_h(\alpha)/I_h(\alpha=0)$) on the angle of attack α of the probe opening in various flow regions of a standard separation nozzle run on a He/SF$_6$ mixture.
(a) Flow region with low gradients of the speed ratio.
(b) Flow region with high gradients of the speed ratio normal to the direction of flow.
(c) Flow region with high gradients of the speed ratio in the direction of flow (strong deceleration).
The curves plotted in the diagrams were fitted to the measured values in (a) on the basis of a local Maxwellian distribution, in (b) and (c) on the assumption of a bimodal velocity distribution

components of a mixture; some preliminary results indicate that, near the deflection wall, the temperature of the heavy component of the mixture considerably exceeds that of the auxiliary gas /41/. Much more information is furnished by the free molecular probe technique with mass spectrometric gas analysis, as will be shown below.

Figure 5.10 represents some typical results of free molecular probe measurements performed in various flow regions of a separation nozzle operated with a He/SF$_6$ mixture. In each case, the normalized ion current of SF$_6$ is plotted as a function of the angle of attack of the probe opening for flow regions with low velocity gradients (Fig.5.10a), with high velocity gradients normal to the direction of flow (Fig.5.10b), and with strong deceleration of the flow (Fig.5.10c). For the light auxiliary gas, qualitatively similar ion current curves result.

At low local velocity gradients (Fig.5.10a), the ion current has a maximum with the opening of the probe facing upstream ($\alpha = 0°$) and a minimum with the opening of the probe facing downstream ($\alpha = \pm 180°$); the ion current curve is symmetrical relative to the maximum. The curve plotted in the diagram was fitted to the measured values of the ion current under the assumption of the local velocity distribution corresponding to a Maxwellian distribution. The good agreement between the measured points and the calculated curve suggests that the heavy component in this flow region is approximately in a state of local thermodynamic equilibrium.

In flow regions with high velocity gradients normal to the direction of flow, which prevail, e.g., at the deflection wall, the ion current curve has an asymmetric shape (Fig.5.10b). The ion current changes greatly in the angular sector in which the probe opening faces regions with high speed ratio ($\alpha \cong +90°$), less so if the opening faces regions with a low speed ratio, i.e., towards the deflection wall ($\alpha \cong -90°$). The curve plotted through the measured points was calculated on the basis of the model assumption that the local velocity distribution may be taken as a superposition of two Maxwellian half-range distributions. The molecules assigned to the two half-range distributions pass through a molar stream surface of the respective component of the mixture in opposite directions. Their number densities and their most probable thermal velocities are determined by the condition that the net flow rate through a stream surface by definition disappears. Since only molecules can enter the probe from the region in front of the probe opening, the two half-range distributions make different contributions to the ion current, depending on the orientation of the probe opening. One contribution always disappears if the probe opening is oriented parallel to the direction of flow. Consequently, the speed ratios to be assigned to the two half-range distributions can be determined from the gradients of the ion current curve at the angles $\alpha = +90°$ and $\alpha = -90°$, respectively, as will be evident from (5.6).

The origins of such bimodal velocity distribution composed of two half-range distributions become evident if a flow is considered with a velocity slip at a fixed wall, which is typical of the range of Knudsen numbers encountered in the separation nozzle flow. In the immediate vicinity of the wall the local velocity distribution is composed of the half-range distribution of the molecules impinging upon the wall at a finite flow velocity and the half-range distribution of the molecules reflected at the wall; the flow velocity of the reflected molecules disappears in the diffuse reflection normally encountered. Consequently, the term "half-range distribution" implies that the respective molecules have either no velocity component in the direction towards the wall or no velocity component in the opposite direction. A line of thought analogous to this concept can be developed by looking at the temperature jump at the wall as detected by free molecular temperature probe measurements (Sect.5.3.2).

In Figure 5.11, speed ratio profiles of the heavy component of a He/C_7F_{14} mixture are shown at various angles of deflection, which are assigned to the half-range distributions. The experimental conditions are identical to those applying to the measurements shown in Fig.5.4, in which only the effective value of the speed ratio had been plotted as calculated from the ratio of ion currents at the angles $\alpha = 0$ and $\alpha = 90°$ in accordance with (5.7). The effective value approximately agrees with the arithmetic mean of the different speed ratios of the two half-range distributions. It is evident from Fig.5.11 that the relative difference between the two half-

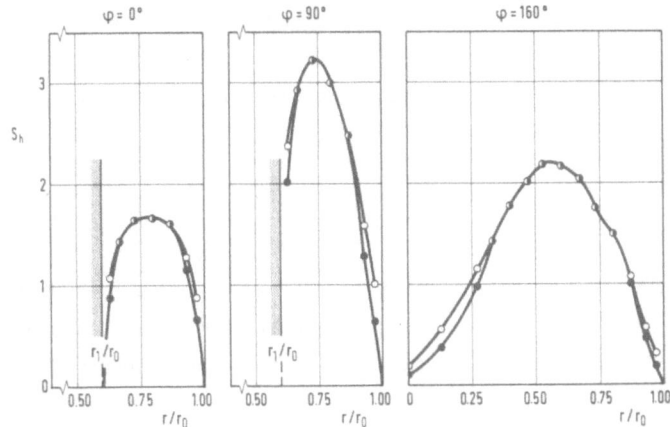

<u>Fig.5.11.</u> Results of free molecular probe measurements in a He/C_7F_{14} model gas mixture of the half-range speed ratio profiles of the heavy component at various angles of deflection ϕ of a standard separation nozzle. The hollow measured points apply to the molecules coming from regions with a high speed ratio, the solid measured points to the molecules coming from the solid walls or the free boundary of the jet (for the experimental conditions, see Fig.5.4)

range speed ratios increases if the distance from the solid wall or the free boundary of the jet ($\phi = 160°$) decreases. At the position of the probe closest to the wall, the speed ratio of the molecules approaching the wall is about 50% higher than the speed ratio of the molecules reflected at the wall and partly swept along again by the flow.[9] Along the free boundary of the jet there are also marked differences in the half-range speed ratios. In accordance with the model concept developed here, these differences can be explained by the penetration of molecules from the stagnant gas into the flow.

The ion current curve shown in Fig.5.10c, which has a secondary peak for the probe opening facing downstream ($\alpha = 180°$), is typical of strong deceleration of the flow within a few mean free paths, occurring in the stagnation zone upstream from the skimmer. The curve plotted through the measured points was fitted again on the basis of a bimodal velocity distribution; in analogy to the Mott-Smith mod-

---

[9] For the sake of completeness, it should be pointed out that the change in position of the probe opening, which is associated with the rotation of the probe, also contributes slightly to the asymmetry of the ion current curve. However, this change in position of ±0.2 mm is reflected only in a distortion, which amounts to less than half the thickness of the curve lines shown in Fig.5.11.

el of the distribution function in shock waves.[10] it was assumed that two flows with different speed ratios, S_{i1} and S_{i2}, and Maxwellian distributions, f_{i1} and f_{i2}, and with different number densities, ν_{i1} and ν_{i2}, are superposed locally. Accordingly, the local velocity distribution f_i follows from the relation

$$\nu_i f_i = \nu_{i1} f_{i1} + \nu_{i2} f_{i2} \qquad (5.16)$$

where ν_i is the local number density of the component i. This model concept is backed by the fact that in free molecular probe measurements in normal shock waves, a qualitatively similar behavior of the ion current was found to that in the stagnation zone upstream from the skimmer /58/.

It follows from the molecular probe measurements that the speed ratios of the light auxiliary gas and the heavy component of the mixture do not behave like the square roots of the molecular masses in those flow regions in which the gas mixture is greatly accelerated or decelerated. For instance, it is observed that rapid acceleration of the mixture in a strongly convergent nozzle causes the speed ratio of the light auxiliary gas normalized to the square root of the molecular mass, i.e., the quantity $S_a/\sqrt{M_a}$, to be clearly higher than the $S_h/\sqrt{M_h}$ value measured for the heavy component of the mixture at the same position in the flow field /66/. The dependence of the ion current, however, on the angle of attack of the probe opening for both components of the mixture corresponds to the curve to be expected for a local Maxwellian distribution. This fact can be explained by the light auxiliary gas reacting more quickly to changes in pressure, because of its higher thermal velocity, than the heavy component of the mixture, which results in a local velocity slip between the components. In the same connection it was found that the flow rate of an H_2/UF_6 mixture through a strongly convergent nozzle of the type used, e.g., in the opposed jet separation nozzles described in Sect.8.3, is anomalously high /66/. At Reynolds numbers around 100 and high expansion ratios, the actual mass flow rate $\dot{m}$ of an H_2/UF_6 mixture is even higher than the maximum isentropic (critical) flow rate $\dot{m}_{is}$ of a uniform gas whose molecular weight and specific heat ratio correspond to the respective mean values of the mixture. Consequently, the mixture, whose components are coupled only weakly in fast changes of state because of the high relative mass dif-

---

[10] Mott-Smith's model starts from the assumption that the velocity distributions on both sides of the shock wave extend right into the shock wave and coexist with different number densities. For this reason, the local velocity distribution within the shock wave is set up as a superposition of two Maxwellian distributions, f_1 and f_2, the respective number densities, ν_1 and ν_2, disappearing at the front and rear ends of the shock, respectively /65/.

ference and the resulting ineffective exchange of translational energy, has a higher
sonic velocity than the corresponding uniform gas.

When interpreting free molecular probe measurements with respect to non-equilibrium
between the components of a mixture, it should be taken into account that only the
speed ratios, not the absolute values of flow velocities and temperatures of the individual components can be determined by this technique. Therefore, it is impossible
to decide, without additional assumptions or information from independent measurements, whether a difference observed locally between the $S_a/\sqrt{M_a}$ and $S_h/\sqrt{M_h}$ ratios is
due to a velocity slip between the components of the mixture or to a difference in
the translational thermal energies of the light and heavy molecular species. Since
a velocity slip induces a difference in temperatures between the components of the
mixture /67/, one must assume on principle that there are differences both in temperature and in velocity if large differences are found locally between $S_a/\sqrt{M_a}$ and
$S_h/\sqrt{M_h}$.

5.4 Entropy Generation in the Separation Nozzle Flow

The specific energy consumption of the separation nozzle process is determined mainly by the increase in the entropy of the process gas, which is associated with the
expansion from the nozzle inlet pressure to the suction pressures of the light and
heavy fractions. Since this expansion occurs without work being performed and at a
constant ambient temperature, the kinetic energy of the partial streams extracted
from the separation element is dissipated practically completely. The reduction in
entropy brought about by the partial separation of the auxiliary gas and UF_6 is can-
celled by the remixing of the light and heavy fractions in the cascade circuit. Consequently, the increase in entropy of the process gas corresponds to that of isenthalpic throttling from the inlet pressure p_0 to the intake pressure of the compressor of a separation stage. If, in a simplified approach, the intake pressure of the
compressor is set equal to the suction pressure p' of the light fraction, the entropy increment associated with the generation of the separation nozzle flow is identical with the entropy increment Δs_{max} as defined in (5.5). The change in state of
the process gas characterized by this throttling process is reversed in the compressor and the gas cooler of the separation stage, which would require compression work
equivalent to $T_0 \cdot \Delta s_{max}$ in the case of ideal isothermal compression. In a commercial
separation stage, as mentioned in Sect.3, additional losses occur as a result of the
nonideal compression and the pressure drop in the cooler and the pipelines. The specific energy consumption of a whole separation plant further includes the reduction

in separation capacity because of mixing losses in the cascade, losses associated with energy conversion, and the power requirement of the auxiliary systems.

The entropy increment Δs_{max} associated with the generation of the separation nozzle flow is best split into two terms, one term characterizing the entropy increment Δs_1 of the process gas up to the splitting of the flow by the skimmer, the other term denoting the entropy increment Δs_2 in the partial streams withdrawn from the separation element up to the intake of the compressor. In accordance with the free molecular probe measurements outlined in Sect.5.3.2, the speed ratio of the gaseous mixture in the separation nozzle is only roughly half the value resulting in the ideal case of isentropic expansion. At the operating conditions typical of the separation nozzle process, the entropy increment Δs_1 of the gaseous mixture in the separation nozzle consequently has an essential influence on the radial pressure differences in the flow and thus on the elementary effect of isotope separation.

The entropy increment Δs_1 was investigated in a standard separation nozzle over a broad range of operating conditions by measuring the pressure losses, as explained in Sect.5.3.1, using various uniform gases and an H_2/C_7F_{14} model gas mixture /57,64/. In Fig.5.12, the relative entropy increment $\Delta s_{rel} = \Delta s_1/\Delta s_{max}$ determined in accordance with (5.14) from the measured pressure losses, has been plotted for the uniform gases and the H_2/C_7F_{14} mixture as a function of the Reynolds number of the separation nozzle flow. It is seen that Δs_{rel} rises continuously with decreasing Reynolds number, which indicates the growing influence of viscous effects upon the separation nozzle flow. In the range of Reynolds numbers optimal for uranium isotope separation (Re = 50 to 150), Δs_{rel} is between 0.5 and 0.6 for uniform gases, i.e., the dissipative losses in the separation nozzle are about as high as the losses in the gas streams extracted from the separation element (Sect.5.2).

When comparing the Δs_{rel} values of the mixture and the uniform gases, it must be taken into account that the pressure losses give only an approximate value of the

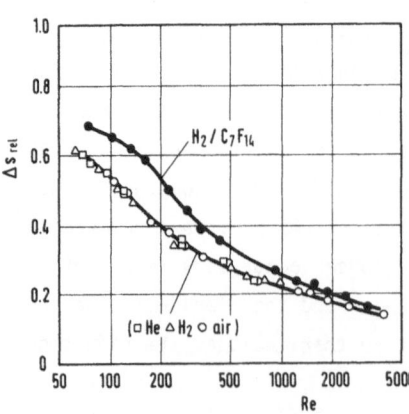

Fig.5.12. Influence of the Reynolds number Re on the relative entropy increment Δs_{rel}, for an H_2/C_7F_{14} mixture and various uniform gases at the end of deflection in a standard separation nozzle (95 mole % of H_2/5 mole % of C_7F_{14}, $p_0/p' = 4$)

entropy increment in the separation nozzle flow, since the reduction in entropy due to the partial separation of the mixture components and also deviations of the local stagnation temperature T_s from the reservoir temperature T_0 are neglected. Nevertheless, it is evident from Fig.5.12 that the relative entropy increment of the H_2/C_7F_{14} mixture, especially in the range of Reynolds numbers optimal for separation, is higher than the entropy increment of uniform gases. The first reason to be considered in this connection is that in the diffusion process the kinetic energy of the relative motion of the components of the mixture is converted into internal energy and, consequently, entropy is generated /67/. Moreover, the radial transport of the heavy component raises the viscosity and the velocity gradients in the flow region at the deflection wall, which further enhances dissipative losses.

In Figure 5.13, the difference has been plotted between the relative entropy increment $\Delta s_{rel,m}$ of H_2/C_7F_{14} mixtures and the relative entropy increment $\Delta s_{rel,p}$ of uniform gases as a function of the molar fraction of C_7F_{14} at a constant expansion ratio ($p_0/p' = 4$) and a constant Reynolds number (Re = 150) of the separation nozzle flow. This difference, which may be considered as a measure of the additional pressure losses associated with the separation of the components of the mixture, passes through a maximum at a C_7F_{14} molar fraction of about 0.02. For other Reynolds numbers and expansion ratios, the relative entropy increment is also most pronounced at molar fractions between $N_h = 0.01$ and $N_h = 0.04$, as was found in additional measurements /64/. Since there is no major difference with respect to flow and separation behavior between H_2/UF_6 and H_2/C_7F_{14} mixtures, it may be concluded that the pressure losses associated with the separation of the mixture components will be most pro-

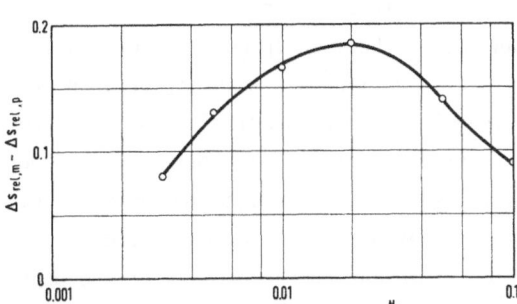

Fig.5.13. Influence of the molar fraction N_h of the heavy component on the relative entropy increment $\Delta s_{rel,m}$ of H_2/C_7F_{14} mixtures at a constant Reynolds number of the separation nozzle flow. The relative entropy increment of uniform gases at the same Reynolds number (Re = 150, $p_0/p' = 4$) is characterized by $\Delta s_{rel,p}$

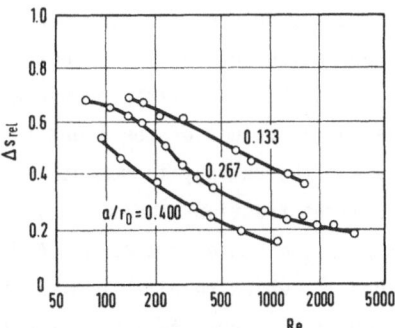

Fig.5.14. Relative entropy increment Δs_{rel} versus Reynolds number Re for various normalized widths a/r_0 of the standard separation nozzle (95 mole % of H_2/5 mole % of C_7F_{14}, $p_0/p' = 4$)

58

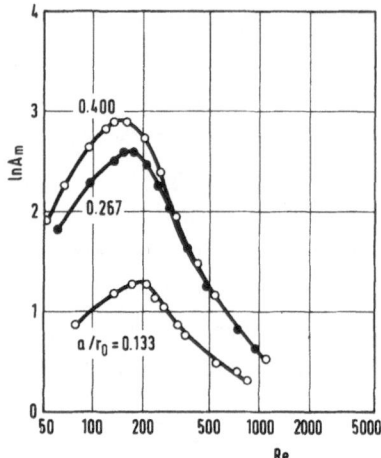

Fig.5.15. Separation factor A_m of an H_2/C_7F_{14} mixture versus Reynolds number Re for various normalized widths a/r_0 of the standard separation nozzle (95 mole % of H_2/5 mole % of C_7F_{14}, $p_0/p' = 4$; at a given Reynolds number, the values of the C_7F_{14} cut were made identical for all nozzle widths by adjusting the position of the skimmer)

nounced at the composition of the process gas customarily used for uranium isotope separation.

It is evident from Fig.5.14 that the dissipative losses in the separation nozzle depend substantially on the ratio of the nozzle width, $a = r_0 - r_1$, to the radius of curvature r_0 of the deflection wall. If the a/r_0 ratio is increased from 1:7.5 to 1:2.5, the relative entropy increment drops by some 30% in the range of Reynolds numbers optimal for isotope separation. The reduction of Δs_{rel} as a/r_0 is being increased can be explained by the reduction in the ratio of the length to the width of the jet, which results in a decrease of the frictional drag of the nozzle at a given Reynolds number. Because of the higher speed ratio the separation factor of the mixture increases with a/r_0, as is evident from Fig.5.15.

At the low cuts of the heavy component of the mixture, which are optimal for isotope separation ($\theta_s \cong 0.2$; cf. Sects.4.1.2 and 6.3), the mean speed ratio of the mixture $\bar{S}_m$ and the mean static pressure of the heavy fraction are much higher than the corresponding values of the light fraction (Fig.5.6). Correspondingly, the pressure losses and the entropy increment in the heavy fraction stream are clearly lower than in the light fraction stream. In a small radial section, the stagnation pressure of the H_2/C_7F_{14} mixture at the end of deflection may even rise above the nozzle inlet pressure p_0, which results from the high speed ratio and number density of the heavy component in a narrow region of the flow /64/.

Because of the high mean stagnation pressure of the flow close to the deflection wall, the suction pressure p" of the heavy fraction stream may be raised considerably above the suction pressure p' of the light fraction stream without causing the separation characteristics of the flow upstream from the skimmer to deteriorate.[11] If the expansion ratio of the heavy fraction is decreased, the total entropy increment associated with the generation of the separation nozzle flow is lower than the Δs_{max} value given by (5.5), for which identical suction pressures were assumed for the light and the heavy fractions. In practice, however, this possibility of reducing the entropy increment and reducing the specific energy consumption has not yet been exploited; the advantage of approximately 10% less compression work is offset by the disadvantage of a more complicated design of the compressors of the separation stage, which would have to process two gas streams of different compositions at different pressure levels. One attractive practical possibility to use the high stagnation pressure of the heavy fraction stream, however, would be to feed the heavy fraction of a first separation nozzle into another separation nozzle directly coupled for subsequent separation. In this way, the separative capacity of the whole system increases without additional power consumption of the compressor /29,68,69/ (cf. Sect.8.2).

An isentropic pressure recovery of the light fraction stream, which contains some 75% of the total molar throughput of the separation nozzle, would decrease the compression work required to generate the separation nozzle flow by some 25%. However, it is not possible in principle to achieve sizable pressure recovery in *one single separation nozzle,* because the low Reynolds number of the light fraction stream makes any major conversion of dynamic pressure into static pressure impossible in a conventional diffusor. However, a considerable pressure recovery would be attainable if the light fraction streams of a *large number of separation nozzle systems* in the form of many adjacent parallel gas jets were combined so as to constitute a flow of a high Reynolds number immediately after splitting the flow by the skimmer. In such arrangements, there may be effective conversion of dynamic pressure into static pressure even at low Reynolds numbers of the single streams, as can be shown by a simple momentum balance /64/. In addition, the dynamic pressure of the high Reynolds number flow generated by the multitude of individual low Reynolds number streams can be effectively recovered in a conventional diffusor. However, no practical use has so far been made of the possibility of pressure recovery in the light fraction, because major design and fabrication difficulties in the required close bundling of the light fraction streams of a multitude of separation nozzles must first be overcome.

---

[11] Such "stagnation of the heavy fraction" normally even has a positive effect on the separation characteristics of the flow field. It causes the static pressure at the deflection wall and, consequently, the radial differential pressure to rise, which corresponds to a higher intensity of the centrifugal field in accordance with (5.15) (Sect.6.2).

6. Influence of the Operating Conditions on Isotope Separation and Specific Expenditure

The operational state of a standard separation nozzle of a given geometry can essentially be regarded as defined if the following conditions have been fixed:

- inlet pressure p_0,
- expansion ratio p_0/p' of the light fraction,
- stagnation ratio p''/p',
- UF$_6$ molar fraction N_u^0 of the feed gas,
- UF$_6$ cut θ_u,
- operating temperature T_0,
- type of auxiliary gas.

In the practical application of the separation nozzle method, the fixing of these operating conditions is based on the requirement to minimize the technical expenditure for a given enrichment problem. The most important individual contributions to this technical expenditure may be characterized in a standardized physical form by the specific parameters explained in Sect.3. As a matter of fact, the specific energy consumption, the specific suction volume, and the specific slit length attain their minimum values under widely different operating conditions, and it is impossible to indicate any simple and clearcut relations among these quantities. Therefore, it is a factor of fundamental importance in designing technical facilities to determine the separation properties of the nozzle over a sufficiently broad range of operating conditions.

In this chapter, the influence of the operating conditions on the separation effect and the specific parameters is described on the basis of the analysis of flow and separation processes and on the basis of separation experiments with UF$_6$ and of free molecular probe measurements with model gases. The experimental results were obtained within a long-term development partly using separation nozzle systems of different geometries. Therefore, quantitative consistency of all experimental data is not always assured. In addition, many cases will have to be based on interpolated data because it was not possible, for experimental reasons, to vary in the experiments only one operating condition while keeping constant all others. The experimental curves shown below therefore should frequently be regarded only as cross sections through a multi-dimensional, complex correlation between the operating conditions and the characteristic parameters of a separation nozzle.

6.1 Inlet Pressure and Expansion Ratio

Figure 6.1 shows a few typical measurements of the dependence of the elementary effect ε_A of isotope separation on the inlet pressure p_0 at various expansion ratios p_0/p'; the separation experiments were performed on an H_2/UF_6 mixture at a constant uranium cut ($\theta_u = 0.25$) in a standard separation nozzle. It is seen that ε_A passes through a maximum as a function of p_0. The maximum is shifted towards higher inlet pressure with increasing expansion ratio and rises in its absolute value.

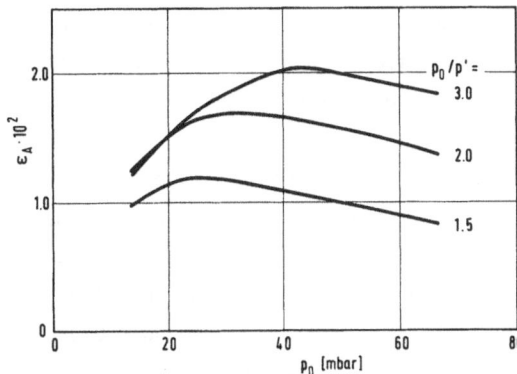

Fig.6.1. Elementary effect ε_A of isotope separation versus inlet pressure p_0 for various expansion ratios p_0/p'. Results of separation experiments using an H_2/UF mixture in a standard separation nozzle; $N_u^0 = 0.03$, $\theta_u = 0.25$, $p' = p''$, $T_0 = 296$ K $r_0 = 0.75$ mm

This finding can easily be interpreted by looking at the dependence on the inlet pressure and the expansion ratio of the main parameters affecting the development of the isotope separation, which are the UF_6 speed ratio and the Knudsen number. The UF_6 speed ratio rises with increasing expansion ratio in accordance with the increase in the Mach number of the mixture and with increasing inlet pressure, since the Reynolds number increases and the influence of viscous effects is reduced. The Knudsen number of the separation nozzle flow changes roughly in inverse proportion to the inlet pressure, and increasing the expansion ratio increases the Knudsen number. As was shown in Sect.4.2.3, increasing the UF_6 speed ratio leads to an increase in the intermediate maximum of the elementary effect of isotope separation and, at constant Knudsen number of the flow, to a more rapid separation; increasing the Knudsen number at a constant UF_6 speed ratio merely produces more rapid separation, i.e., the intermediate maximum of the elementary effect is shifted towards smaller angles of deflection (4.24).

The elementary effect ε_A of isotope separation accordingly must first increase with p_0 at low inlet pressure because, on the one hand, the UF_6 speed ratio increases with p_0 and, on the other hand, the intermediate maximum of ε_A is shifted further downstream to the skimmer. At higher inlet pressures, viscous effects are less important and the speed ratio becomes independent of p_0; ε_A must decrease again with increasing p_0 because, as a result of the decreasing Knudsen number, separation proceeds more slowly and the intermediate maximum of isotope separation can no longer be attained in the range of deflection limited by the skimmer position.

Since increasing the expansion ratio at a constant inlet pressure causes the maximum of isotope separation to be shifted towards smaller angles of deflection, the elementary effect of isotope separation present at the skimmer may even decrease with increasing expansion ratio in the range of low inlet pressures (Fig.4.11). The acceleration of isotope separation, which is due above all to the increase in the UF_6 speed ratio, can be compensated by raising the inlet pressure, so that the maximum of isotope separation is shifted downstream again towards the skimmer. Therefore, the inlet pressure p_0^+ optimal for separation, i.e., the pressure at which the maximum of isotope separation is attained right at the skimmer, rises with increasing expansion ratio.[12]

Figure 6.2 shows a few typical results of separation experiments in the standard separation nozzle, in which the elementary effect ε_A of isotope separation and the throughput of the mixture, L_m, per unit length of the nozzle slit were determined over a wide range of inlet pressures p_0 and expansion ratios p_0/p'. The so-called skimmer distance, i.e., the width of the channel formed by the skimmer and the deflection wall, was always set so that a UF_6 cut of $\theta_u = 0.25$ was the result. The contour lines of ε_A in the p_0-p_0/p' field exemplify the dependence discussed above of the separation effect on the inlet pressure and the expansion ratio, and it becomes particularly evident how the inlet pressure p_0^+ optimal for separation increases with the expansion ratio. The throughput of the mixture, L_m, is roughly proportional to the inlet pressure in the range shown here. Unlike the known throughput behavior of an isentropic, onedimensional continuum flow, in which the throughput through a nozzle no longer changes when the critical expansion ratio has been reached (p_0/p_{crit} =

---

[12] The plus symbol always characterizes the value of a certain operating parameter at which the elementary effect ε_A of isotope separation reaches a maximum if only this parameter is varied, whereas all other operating conditions remain constant. In addition to p_0^+, corresponding "optimum" values can also be defined for the Reynolds number and the Knudsen number or for the expansion ratios of the light and the heavy fractions. In general, the optimum value of a given operating parameter changes if other operating parameters are changed. For instance, p_0^+ rises with increasing expansion ratio or decreasing UF_6 molar fraction (Sect. 4.2.3).

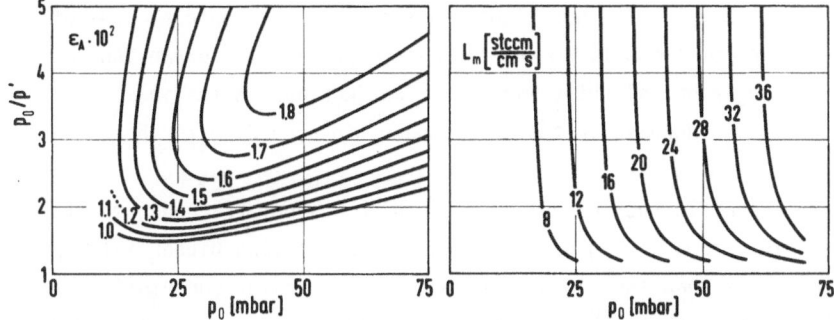

Fig.6.2. Influence of the inlet pressure p_0 and the expansion ratio p_0/p' on the elementary effect ε_A of isotope separation and the throughput L_m of the gaseous mixture. Results of separation experiments using an H_2/UF_6 mixture in a standard separation nozzle; $N_u^0 = 0.03$, $\theta_u = 0.25$, $p' = p''$, $T_0 = 296$ K, $r_0 = 0.75$ mm

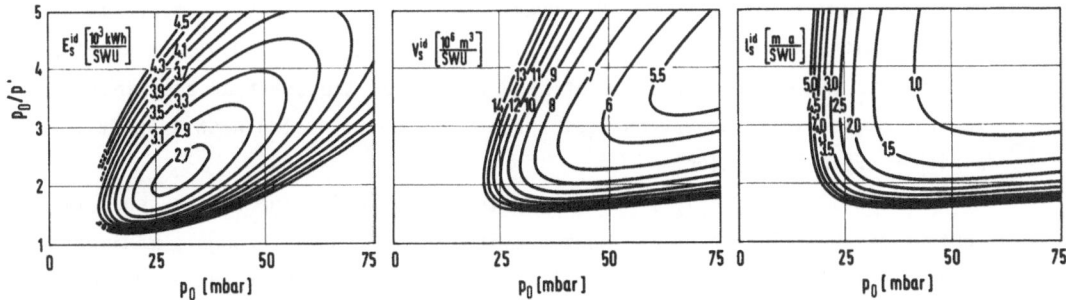

Fig.6.3. Influence of the inlet pressure p_0 and the expansion ratio p_0/p' on the specific energy consumption E_s^{id}, the specific suction volume V_s^{id}, and the specific slit length l_s^{id}. Results of separation experiments using an H_2/UF_6 mixture in a standard separation nozzle; for the operating conditions, see Fig.6.2

1.9 for $N_u^0 = 0.03$), L_m in the curved viscous flow clearly continues to increase up to expansion ratios $p_0/p' \lesssim 3$ and gradually converges against a limit dependent only on the inlet pressure.

The relationship shown in Fig.6.2 between the quantities ε_A, L_m, p_0 and p_0/p' can be used at a given UF_6 cut θ_u and UF_6 molar fraction N_u^0 to determine directly the dependence of the specific process parameters on the inlet pressure and the expansion ratio. Figure 6.3 shows the corresponding contour lines in the p_0-p_0/p' field for the specific energy consumption E_s^{id}, the specific suction volume V_s^{id}, and the specific slit length l_s^{id}, all of which were calculated neglecting the losses unavoidable in a technical plant (3.5-7).

The specific energy consumption E_s^{id} in the separation nozzle system studied here reaches a minimum at an expansion ratio of about 2.2 and at an inlet pressure of approximately 30 mbar[13]. The positions and shapes of the contour lines of E_s^{id} reflect the variation of the optimum inlet pressure p_0^+ with the expansion ratio. It is seen that E_s^{id} changes but slightly with p_0 if an increase in the inlet pressure is associated with an increase in the expansion ratio according to the change of p_0^+ with p_0/p'.

The minimum of the specific suction volume V_s^{id}, which was not reached completely in the contour line plot shown in Fig.6.3, is always found at much higher inlet pressures and expansion ratios than the minimum of the specific energy consumption. This situation is due to the fact that E_s^{id} depends on the expansion ratio p_0/p', while V_s^{id} is inversely proportional to the absolute value of the suction pressure p'. Since p' increases linearly with p_0 at a constant expansion ratio, and ε_A changes only weakly as a function of p_0 in the range of optimum inlet pressures, V_s^{id} initially continues to decrease when p_0^+ is exceeded, while E_s^{id} passes through a minimum at the optimum inlet pressure. Moreover, the separation effect ε_A rises to a higher level with the expansion ratio at high inlet pressures than at the inlet pressure leading to the absolute minimum of E_s^{id}. Therefore, the shift of the minimum of V_s^{id} towards higher inlet pressures must necessarily be connected with a shift towards higher expansion ratios.

The specific slit length l_s^{id}, defined as the reciprocal value of the separative work output per unit length of the nozzle slit, does not pass through a minimum in the p_0-p_0/p' field. It converges against a lower limit at high expansion ratios, which depends only on the inlet pressure. This behavior can be explained by the fact that the effective expansion ratio of the gaseous mixture and, hence, its Mach number is limited in accordance with the given shape of the nozzle and the suction channel of the light fraction. The dependence on inlet pressure of l_s^{id} is determined by the fact that the throughput through the nozzle increases approximately proportional to p_0, whereas the separation effect increases with p_0 at low inlet pressures, reaches a maximum at $p_0 = p_0^+$, and decreases with increasing p_0 at high inlet pressures. Therefore, l_s^{id} decreases with p_0 in the range of low inlet pressures; it continues to decrease if p_0^+ is slightly exceeded, since ε_A changes little in this pressure range. However, as the inlet pressure is raised further, ε_A gradually changes inversely proportional to p_0, because the separating pressure diffusion stream becomes independent of pressure at high Reynolds numbers, while the az-

---

[13] The characteristic dimensions of the separation nozzle systems to be installed in technical plants are smaller by a factor of 15 than those of the laboratory separation nozzle used for the separation experiments; consequently, the inlet pressure for minimal specific energy consumption is higher by that factor (cf. Sect.4.2.3).

imuthal transport stream increases linearly with p_0. The separative work output, in which ε_A is a quadratic term, while the throughput proportional to p_0 is only a linear term, decreases inversely proportional to p_0 at very high inlet pressures. Accordingly, the specific slit length rises linearly with p_0 in this pressure range.

The specific energy consumption and the specific slit length are independent of the characteristic nozzle dimensions which, to simplify matters, have so far been characterized by the radius of deflection r_0. The specific suction volume, however, is proportional to those dimensions and they are made as small as possible in the practical application of the separation nozzle process. The reduction in separation nozzle dimensions technically feasible is determined mainly by the tolerances of the skimmer distance f, because major variations in this characteristic dimension will result in significant losses in separative capacity.

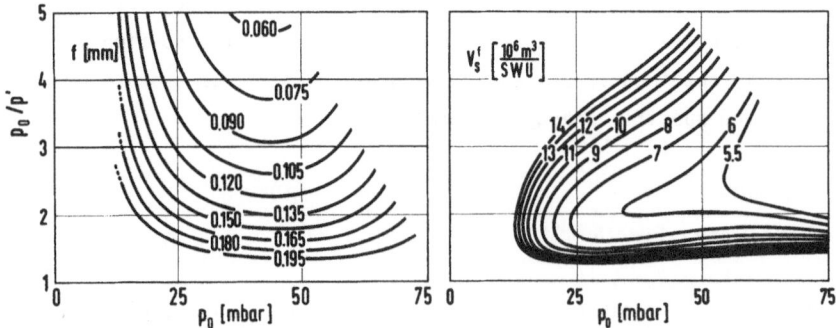

Fig.6.4. Influence of the inlet pressure p_0 and the expansion ratio p_0/p' on the skimmer distance f, for a constant UF_6 cut of 0.25 and on the specific suction volume V_s^f, determined for a constant skimmer distance of 0.1 mm. Results of separation experiments using an H_2/UF_6 mixture in a standard separation nozzle; for the operating conditions, see Fig.6.2

It is evident from the measurements shown in Fig.6.4 that the skimmer distance f required to achieve a certain UF_6 cut θ_u decreases continuously with increasing expansion ratio and passes through a minimum as a function of the inlet pressure. The change in f (θ_u = const) with p_0/p' is due to the fact that increasing the expansion ratio causes the separation factor of the mixture to rise as a result of the rising Mach number of the mixture; consequently, the UF_6 is forced more strongly towards the deflection wall and the skimmer must be positioned closer to the deflection wall in order to retain a given θ_u value. The dependence on the inlet pressure of f (θ_u = const), on the one hand, is determined by the separation factor of the mixture passing through a maximum as p_0 is varied. On the other hand, the whole

mixture is forced more strongly towards the deflection wall with increasing inlet pressure, as the Reynolds number and, hence, the flow velocity increases.

If the specific suction volume is to be determined in the light of practical aspects, the importance of the skimmer distance to the feasible reduction of the nozzle dimensions must be taken into account. Therefore, it is reasonable to presume a fixed value f_0 for the skimmer distance and substitute V_s^{id} by the quantity

$$V_s^f = V_s^{id} \cdot f_0/f \quad . \tag{6.1}$$

The contour lines of V_s^f shown in Fig.6.4 were calculated for $f_0 = 0.1$ mm. Although the minima of V_s^{id} and V_s^f were not fully determined, it is obvious that the minimum of V_s^f is shifted towards lower expansion ratios and lower inlet pressures compared with the corresponding minimum of V_s^{id}. The consequence is that the minimum of V_s^f is closer to the minimum of the specific energy consumption than is the minimum of V_s^{id}. Assuming a skimmer distance of 0.01 mm, which is typical of the present state of the manufacturing methods for technical-scale separation nozzles, a specific suction volume of about 7×10^5 m^3/SWU is obtained if a standard separation nozzle is operated under the conditions of minimum specific energy consumption.

6.2 Stagnation of the Heavy Fraction

In a separation nozzle plant, a definite value for the UF$_6$ cut θ_u is specified by the cascade circuit. This value should be set as precisely as possible in every separation stage in order to minimize losses in separative work due to mixing of UF$_6$ streams of different isotopic compositions. For fine control of θ_u, the position of the skimmer is set in such a way that θ_u is slightly below the set-value, if the suction pressures of the light and the heavy fractions are identical. By means of a control valve installed in the pipeline of the heavy fraction the gas stream of the heavy fraction is reduced and, correspondingly, that of the light fraction is increased until θ_u assumes the set-value specified by the cascade circuit. This causes the pressure p" in the heavy fraction to rise above the suction pressure of the light fraction.

Figure 6.5 shows the change in the UF$_6$ cut for two typical skimmer distances plotted versus the p"/p' ratio of the suction pressures of the heavy and the light fraction, the so-called stagnation ratio/70/. At the small skimmer distance ($f/r_0 = 0.16$), θ_u and p"/p' rise monotonically with increasing throttling, p"/p' attaining its maximum value when the control valve in the pipeline of the heavy fraction is completely closed ($\theta_u = 1$). At the greater skimmer distance ($f/r_0 = 0.253$), θ_u and

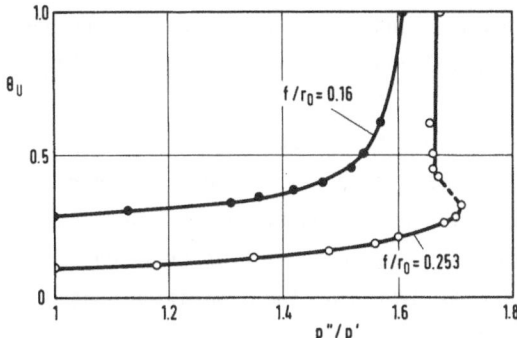

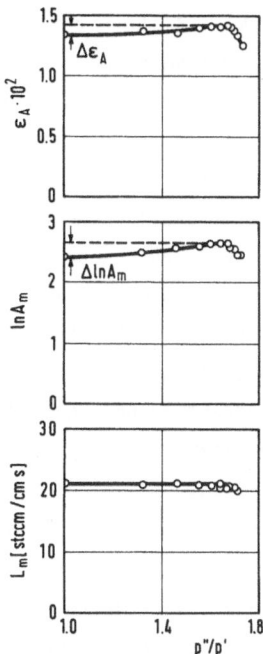

Fig.6.5. Influence of the stagnation ratio p"/p' on the UF$_6$ cut θ_u for two skimmer distances f. Results of separation experiments using an H$_2$/UF$_6$ mixture in a standard separation nozzle; N$_u^0$ = 0.04, p$_0$ = 30 mbar, p$_0$/p' = 2.1, T$_0$ = 296 K, r$_0$ = 0.75 mm

Fig.6.6. Influence of the stagnation ratio p"/p' on the elementary effect ε_A of isotope separation, the separation factor of the mixture A$_m$, and the throughput of the mixture L$_m$. Results of separation experiments using an H$_2$/UF$_6$ mixture in a standard separation nozzle; θ_u = 0.25, N$_u^0$ = 0.042, p$_0$ = 29 mbar, p$_0$/p' = 2.1, T$_0$ = 296 K, r$_0$ = 0.75 mm

p"/p' initially rise continuously as the gas stream of the heavy fraction is throttled. However, further throttling then causes a discontinuous increase in the UF$_6$ cut and a corresponding discontinuous decrease of the stagnation ratio. If throttling is continued further, only θ_u continues to rise, while p"/p' remains constant. The maximum stagnation ratio attainable by throttling increases with the distance f of the skimmer from the deflection wall.

Stagnation of the heavy fraction can significantly improve the separation properties at a given UF$_6$ cut, as is evident from Fig.6.6 /70/. This is a plot of the elementary effect ε_A of isotope separation, the separation factor of the mixture A$_m$, and the throughput of the mixture L$_m$, for an H$_2$/UF$_6$ mixture as a function of the stagnation ratio p"/p'. In these measurements, the skimmer distance was increased step by step and the stagnation ratio was raised by throttling the heavy fraction so as to result in a UF$_6$ cut of θ_u = 0.25. Both ε_A and ln A$_m$ increase with p"/p', passing through a maximum at high stagnation ratios; the relative increase in ε_A is about 5%, the relative increase in ln A$_m$ some 10%. The throughput of the mixture, L$_m$, remains constant within the limits of error until the maximum of ε_A has been reached; thus, stagnation of the heavy fraction can effect an increase by some 10% in the separative work output and a corresponding reduction in the specific expenditure of the process. This improvement in the separation properties is largely independent of the way in which the other operating conditions are chosen; an increase in ε_A by some 5% will result even if N_u, p_0, p_0/p' and θ_u deviate some 30% to 50% from the operating conditions listed in Fig.6.6.

If it is further taken into account that raising the pressure in the heavy fraction reduces the work required to compress the gas expanded in the nozzle, there is even an overall reduction in the specific energy consumption by 20%. However, this additional reduction in specific energy consumption has not yet been utilized in practice, because it would require a more complex design of the separation stage compressors, as has been explained in Sect.5.4.

Influencing the separation characteristics by throttling the gas stream of the heavy fraction has been studied in detail by free molecular probe measurements using He/SF$_6$ mixtures /71/. In those studies it was found that the increases in the elementary effect of isotope separation and in the separation factor of the mixture can be traced back mainly to an increase in the radial differential pressure which, according to (5.15), characterizes the intensity of the separating centrifugal field. Some typical results of these probe measurements are explained in more detail below.

From Fig.6.7 it appears that the speed ratio S_m of the He/SF$_6$ mixture decreases very strongly at the deflection wall with increasing stagnation ratio, the flow being shifted to smaller radii of curvature. At high stagnation ratios, this may even cause the flow to separate from the deflection wall, which can be concluded from the S_m profile which has a point of inflection near the deflection wall for p"/p' = 1.75. Since the stagnation of the heavy fraction preferably affects the flow close to the deflection wall, the static pressure will rise considerably in this region, while

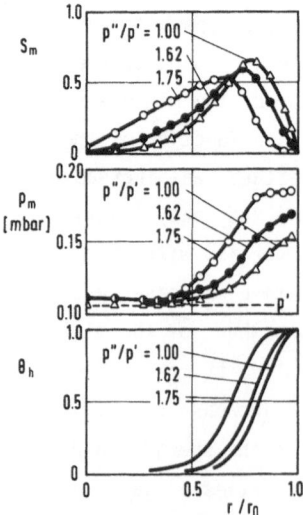

Fig.6.7. Influence of the stagnation ratio p"/p' on the radial profiles of the speed ratio S_m of the mixture, the static pressure p_m of the mixture, and the SF$_6$ cut θ_s in the flow cross section at the angle of deflection of $\phi = 140°$. Results of free molecular probe measurements using a He/SF$_6$ mixture in a standard separation nozzle; $N_h^0 = 0.04$, $p_0 = 0.27$ mbar, $p_0/p' = 2.5$, $r_0 = 150$ mm, $f = 48$ mm; for the nozzle contour, see Fig.5.3

there is only a minor change in pressure at small radii, i.e., the radial differential pressure increases greatly as a function of stagnation. It is seen from the radial profile of θ_s that the skimmer distance must be made larger and larger with increasing stagnation factor, if the flow is to be split at the skimmer at a constant cut of the heavy component of the mixture. The influence exerted on the separation nozzle flow by stagnation is most pronounced in a relatively large region of the flow. If one applies, e.g., the operating conditions leading to a minimum specific energy consumption, stagnation of the heavy fraction changes the flow properties up to an angle of deflection of approximately 90°.

The improvement in isotope and mixture separation connected with stagnation is indicated by the results of free molecular probe measurements shown in Fig.6.8, where the elementary effect of isotope separation and the separation factor of the mixture are plotted versus the SF_6 cut for a flow with equal suction pressures of the light and heavy fractions (p"/p' = 1).[14] It is seen that ϵ_A and ln A_m are clearly higher for all values of θ_s, if the suction pressure p" of the heavy fraction exceeds that of the light fraction.

Any more detailed analysis of the influence of stagnation must take into account that the structure of the flow field in the separation nozzle is clearly changed as a result of changes in the skimmer position. On the one. hand, this is due to the

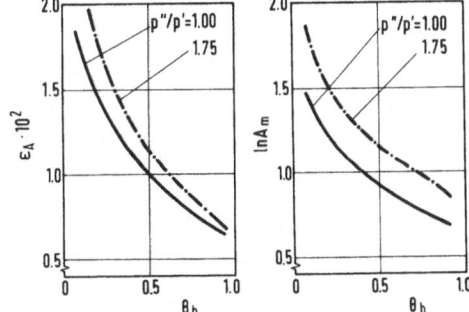

Fig.6.8. Variation of the elementary effect ϵ_A of isotope separation and the separation factor A_m of a He/SF_6 mixture with the SF_6 cut θ_h for equal suction pressures of light and heavy fractions (p"/p' = 1) and for high stagnation of the heavy fraction (p"/p' = 1.75). Results of free molecular probe measurements in a standard separation nozzle in the flow cross section at the angle of deflection of ϕ = 140°; for the operating conditions, see Fig.6.7

[14] In interpreting free molecular probe measurements it must be taken into account that ϵ_A and A_m can be determined for each place in a flow cross section, i.e., for any fictitious cut θ, by integration of the partial fluxes (Sect.5.3.1). In contrast to this situation, the separation experiments using UF_6 include only measurements of the isotopic composition and the composition of the mixture in the partial streams extracted from the separation nozzle at a cut whose value depends on the skimmer position and on the operating conditions. Any change in the cut will affect the flow field, because any change in the skimmer position will have a significant influence on suction conditions, even if all other boundary conditions are kept constant.

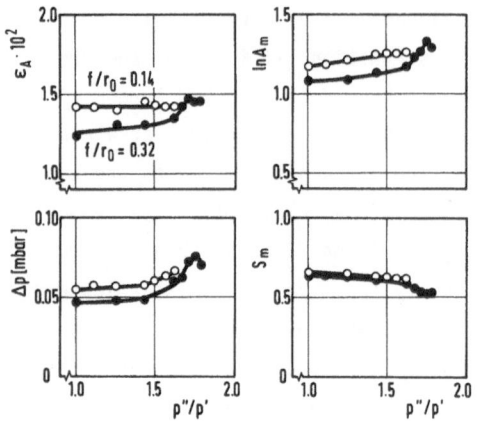

Fig.6.9. Influence of the stagnation ratio p"/p' on the elementary effect ε_A of iso-tope separation, the separation factor A_m of the mixture, the speed ratio S_m of the mixture and the difference between the static pressures Δp at the outer and in-ner boundary of the flow for two skimmer distances, f. The values of ε_A, A_m, and S_m refer to the molar stream surface $\theta_S = 1/3$. Results of free molecular probe measure-ments in a standard separation nozzle in the flow cross section at the angle of de-flection of $\phi = 140°$; for the operating con-ditions, see Fig.6.7

skimmer acting as a solid obstacle in the flow (Fig.5.5), on the other hand, the flow resistance of the suction channels of the light and the heavy fractions changes as a function of the skimmer position.

This situation is exemplified in the probe measurements shown in Fig.6.9, in which the influence of stagnation on flow and separation processes was studied for two different skimmer distances. First of all, it is seen that the elementary ef-fect of isotope separation and the separation factor of the mixture for the SF_6 mo-lar stream surface $\theta_S = 1/3$ are clearly higher for the smaller than for the larger skimmer distances in case of equal suction pressures of the light and the heavy frac-tions (p"/p' = 1). When raising the stagnation ratio up to a maximum dependent on f (Fig.6.5), no increase in ε_A ($\theta_S = 1/3$) is observed at the small skimmer distance, and only a slight one in $\ln A_m$ ($\theta_S = 1/3$). At the large skimmer distance, which enables higher stagnation ratios to be achieved, ε_A and $\ln A_m$ rise above the values attainable at a small skimmer distance shortly before reaching the maximum stagna-tion ratio.

The changes in ε_A and $\ln A_m$ with f and p"/p' are largely parallel to the corre-sponding change in the radial differential pressure Δp, which characterizes the in-tensity of the centrifugal field. Hence, it is obvious to ascribe the effect of stagnation to the parameters influencing Δp. Reducing the skimmer distance, on the one hand, has the consequence that the rising resistance of the suction channel of the heavy fraction causes the flow to be decelerated upstream from the skimmer and the static pressure at the deflection wall to rise. On the other hand, the expan-sion of the heavy fraction downstream from the deflection zone is that much less able to cause a decrease in the static pressure at the deflection wall upstream from the skimmer, the smaller the channel width f. However, if the flow of the heavy fraction is throttled, higher stagnation pressures are attained for the larger skimmer distance (Fig.6.5), because flow regions with higher stagnation pressures and, hence, higher kinetic energies are decelerated. If the heavy fraction is strongly throttled, the static pressure at the deflection wall for large skimmer distances rises above the maximum static pressure for small skimmer distances.

The increase in static pressure at the deflection wall is limited by the flow being separated from the deflection wall; in the separated flow, further throttling of the influx into the heavy fraction does not greatly change the stagnation ratio, as is evident from the measurements shown in Fig.6.5. If the skimmer distance is large, strong throttling of the heavy fraction is even associated with a decrease of Δp. This is due to the fact that an increase in f results in a corresponding decrease in the cross section of the suction channel of the light fraction so that the pressure losses in this channel increase strongly with gas throughput and cause the static pressure at the inner boundary of the jet to rise. Consequently, ε_A and $\ln A_m$ decrease at large skimmer distances and very high stagnation ratios, as can be seen in Figs.6.6,9.

6.3 UF$_6$ Cut

The model calculations of isotope separation in the separation nozzle had shown the elementary effect of isotope separation to grow continuously with decreasing UF$_6$ cut at a given structure of the centrifugal field. However, any change in the cut requiring a change in skimmer position at given operating conditions will influence the structure of the flow field, as has been explained in the previous chapter. Consequently, it must be assumed that the theoretical relationship between ε_A and θ_u may deviate clearly from the experimental relationship, if merely the skimmer distance is varied under otherwise constant operating conditions.[15]

In Fig.6.10, the results are shown for a typical series of measurements of the dependence of the elementary effect ε_A of isotope separation and the separative work output δU on the UF$_6$ cut. The curves plotted in the diagrams correspond to the equations

$$\varepsilon_A = \text{const} \ (\ln \theta_u)/(1-\theta_u) \tag{6.2}$$

and

$$\delta U = \text{const} \ \theta_u(1-\theta_u)\left[(\ln \theta_u)/(1-\theta_u)\right]^2 \quad , \tag{6.3}$$

---

[15] The UF$_6$ cut can be adjusted within certain limits also by a change in operating pressures at a given skimmer position, as has been explained on the basis of Fig. 6.4. However, this possibility is not considered in Sect.6.3.

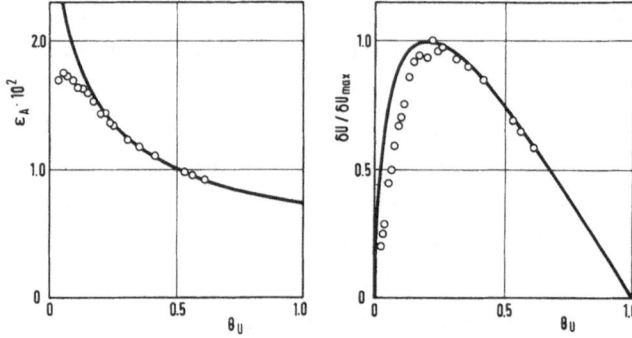

<u>Fig.6.10.</u> Influence of the UF$_6$ cut θ_u on the elementary effect ε_A of isotope separation and on the normalized separative power $\delta U/\delta U_{max}$. The measured values were determined by separation experiments using an H$_2$/UF$_6$ mixture in a standard separation nozzle (N$_u^0$ = 0.04, p$_0$ = 25 mbar, p$_0$/p' = 1.9, p" = p', r$_0$ = 0.75 mm). The solid curves show the dependence on θ_u for equilibrium separation; they were fitted to the measured values at θ_u = 1/3

applying to equilibrium separation and Rayleigh distillation (Sect.4.1.2); they were fitted to the measured values at θ_u = 1/3. It is seen that the experimental correlations $\varepsilon_A(\theta_u)$ and $\delta U(\theta_u)$ are described very well by (6.2,3), respectively, for θ_u > 0.2. However, at lower UF$_6$ cuts the measured values deviate greatly from equilibrium separation and ε_A even decreases with decreasing θ_u.

As explained in Sect.6.2, this result can be traced back to the fact that the centrifugal forces upstream from the skimmer are reduced considerably by the expansion of the heavy fraction at the end of the deflection wall, if the skimmer distance is large and the resultant UF$_6$ cut is small. This explanation is backed by the finding that the deviation of the experimental behavior from equilibrium separation becomes greater if the expansion ratio p$_0$/p' is decreased or if the UF$_6$ molar fraction N$_u^0$ is increased. In both cases, the skimmer distance for a given UF$_6$ cut must be increased and the maxima of ε_A and δU are shifted towards higher values of θ_u.

If the separation nozzle is operated at conditions typical of the technical application of the process, the cut most favorable with respect to the separative work output is around 0.2 (Fig.6.10); accordingly, the specific process parameters E_s^{id}, V_s^{id}, and l_s^{id} pass through a minimum at this θ_u value, if all other operating conditions are kept constant. However, the choice of the cut for a technical separation nozzle cascade is determined not only by the need to find the most favorable specific process parameters, but also by criteria demanding a simple cascade circuit and a small number of separation stages. In the light of these aspects, it is evident that especially the UF$_6$ cuts of θ_u = 1/3 and 1/4 are attractive for technical facilities, since they result in the simplest nonsymmetric cascade circuits for θ_u < 0.5. Operation at θ_u = 1/5 results only in a very slight increase in δU compared with θ_u = 1/4, the drawbacks associated with the more complicated cascade circuit and the higher number of stages at θ_u = 1/5 overcompensating the marginal advantage of the slight increase in separative power. In addition, as mentioned above, the separative work output may reach its maximum at θ_u = 1/4 for low expansion ratios and high UF_6 molar fractions. Operation at θ_u = 1/2, which results in the simplest cascade circuit and the smallest number of separation stages, has the disadvantage of the separative work output being some 25% lower than at θ_u = 1/4. Consequently, a UF_6 cut of θ_u = 1/2 will hardly be attractive for technical separation nozzle plants.

6.4 UF_6 Molar Fraction

In connection with the theoretical description of uranium isotope separation by means of the ternary diffusion equations, it has been shown in Sect.4.2.1 that the elementary effect of isotope separation rises strongly if the UF_6 molar fraction N_u^0 of the process gas mixture is reduced. At the same time, the maximum of isotope separation at a given throughput of the mixture will be shifted towards smaller angles of deflection if the UF_6 molar fraction is decreased. The Knudsen number of the separation nozzle flow optimal for isotope separation thus decreases with decreasing UF_6 molar fraction, if one presumes a fixed angle of deflection of the separation nozzle (Sect.4.2.3).

This behavior is apparent also from the separation experiments shown in Fig.6.11, which had been carried out on H_2/UF_6 mixtures with different UF_6 molar fractions N_u^0 at p_0/p' = 2 and θ_u = 0.25. It is seen that the maximum attainable value of ε_A rises approximately by a factor of 3 when changing from $N_u^0 = 0.1$ to $N_u^0 = 0.01$. At the same time, the inlet pressure p_0^+ optimal for separation rises with decreasing UF_6 molar fraction, i.e., the optimum Knudsen number decreases. The relative change $\Delta\varepsilon_A/\varepsilon_A$ of the separation effect, which is associated with the relative change in the molar fraction $\Delta N_u^0/N_u^0$, decreases with decreasing UF_6 molar fraction N_u^0, for the expansion ratio used here (p_0/p' = 2). When changing from $N_{u_0}^0 = 0.1$ to $N_{u_0}^0 = 0.03$, $\varepsilon_A(p_0=p_0^+)$ clearly rises more than $1/\sqrt{N_u^0}$; when changing from $N_u^0 = 0.03$ to $N_u^0 = 0.01$, it rises clearly less than $1/\sqrt{N_u^0}$. This directly indicates that the specific energy consumption, which is reciprocal to $N_u^0 \cdot \varepsilon_A^2$, must pass through a minimum, at the given expansion ratio, in the range of molar fractions between $N_u^0 = 0.01$ and $N_u^0 = 0.1$.

Figure 6.12 shows the optimum inlet pressure p_0^+ versus the expansion ratio for various UF_6 molar fractions. It is seen that p_0^+ rises steeply with the expansion ratio at low UF_6 molar fractions while, at high UF_6 molar fractions, there is only a slight increase in p_0^+ with p_0/p'. At an expansion ratio of $p_0/p' \cong 2$, roughly identical values for the UF_6 throughput L_u will result at molar fractions between $N_u^0 = 0.01$

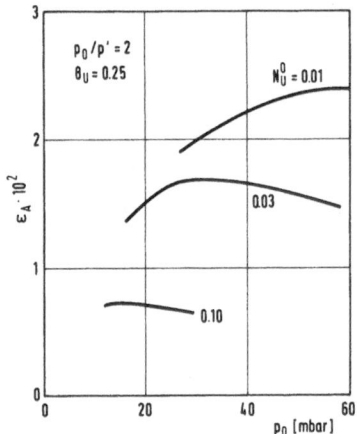

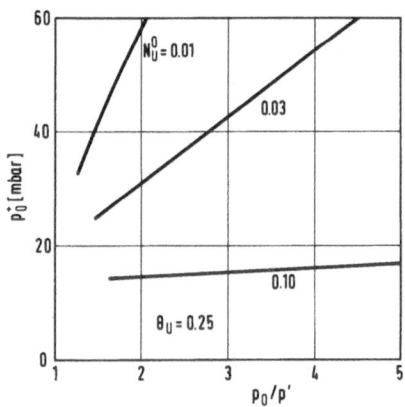

Fig.6.11. Influence of the inlet pressure p_0 on the elementary effect ε_A of isotope separation for various UF_6 molar fractions N_u^0. Results of separation experiments using H_2/UF_6 mixtures in a standard separation nozzle; $\theta_u = 0.25$, $p_0/p' = 2$, $p'' = p'$, $T_0 = 296$ K, $r_0 = 0.75$ mm

Fig.6.12. Variation of the optimum inlet pressure p_0^+ with the expansion ratio p_0/p' for various UF_6 molar fractions N_u^0. Results of separation experiments using H_2/UF_6 mixtures in a standard separation nozzle; $\theta_u = 0.25$; $p'' = p'$, $T_0 = 296$ K, $r_0 = 0.75$ mm

and $N_u^0 = 0.1$, if the separation nozzle is operated at the inlet pressure optimal for isotope separation. Consequently, the reduction of the UF_6 content in the process gas is compensated for by the increase in the optimum inlet pressure and by the increase in the discharge velocity from the nozzle resulting from the reduction in average molecular weight of the H_2/UF_6 mixture. At higher expansion ratios, the UF_6 throughput optimal for separation, L_u^+, even rises with decreasing UF_6 molar fraction. This directly indicates that the specific slit length l_s^{id}, which is reciprocal to $L_u \cdot \varepsilon_A^2$, and the specific suction volume V_s^{id}, which is reciprocal to $p' \cdot N_u^0 \cdot \varepsilon_A^2$, will decrease strongly when changing from $N_u^0 = 0.1$ to $N_u^0 = 0.01$.

Figure 6.13 presents an overview of the influence of the UF_6 molar fraction and the expansion ratio upon the elementary effect of isotope separation and the resultant specific process parameters; the inlet pressure p_0 has been varied according to Fig.6.12 so that $p_0 = p_0^+(N_u^0, p_0/p')$ for all curves shown in Fig.6.13. It is seen that $\varepsilon_A(p_0=p_0^+)$ rises continuously with increasing expansion ratio and decreasing UF_6 molar fraction. At $N_u^0 = 0.03$, the specific energy consumption is some 30% lower than at $N_u^0 = 0.01$ and $N_u^0 = 0.1$. The specific suction volume and the specific slit length decrease by an average factor of 5 when changing from $N_u^0 = 0.1$ to $N_u^0 = 0.03$, and by a factor of 1.5 to 2 when changing from $N_u^0 = 0.03$ to $N_u^0 = 0.01$. According to this behavior it is plausible that in practical application of the separation nozzle method H_2/UF_6 mixtures are used with UF_6 molar fractions of 0.02 to 0.04; this allows an

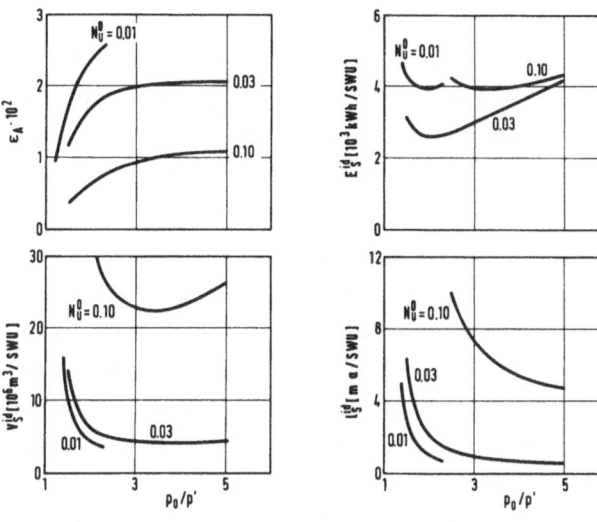

Fig.6.13. Influence of the UF_6 molar fraction N_u^0 and the expansion ratio p_0/p' upon the elementary effect ε_A of isotope separation, the specific energy consumption E_s^{id}, the specific suction volume V_s^{id}, and the specific slit length l_s^{id}; the inlet pressure p_0 has been varied according to Fig.6.12 so that $p_0 = p_0^+(N_u^0, p_0/p')$. Results of separation experiments using H_2/UF_6 mixtures in a standard separation nozzle; $\theta_u = 0.25$, $p'' = p'$, $T_0 = 296$ K, $r_0 = 0.75$ mm

economically optimal compromise to be reached with respect to the different contributions of the various specific process parameters to the costs of separative work (Sect.3).

The dependence of the separation characteristics on the UF_6 molar fraction described here is mainly due to the fact that a change in N_u^0 changes both the UF_6 speed ratio and the D_T/D_m ratio of the ternary diffusion coefficient to the binary diffusion coefficient of the UF_6/auxiliary gas mixture (Sect.4.2.1). Factors less important, though not negligible, are the influence of the UF_6 molar fraction upon the dynamic viscosity of the mixture, which drops by some 40% in H_2/UF_6 mixtures when changing from $N_u^0 = 0.1$ to $N_u^0 = 0.01$, and the entropy production associated with the separation of UF_6 and H_2, which reaches a maximum at $N_u^0 \cong 0.02$ (Sect.5.4, Fig.5.13).

The partial speed ratio of UF_6 rises approximately inversely proportional to the square root of the average molecular weight, if fixed values for the expansion ratio and the Reynolds number are assumed, i.e., if the Mach number of the mixture remains unchanged. On the one hand, the consequence is that any reduction in the UF_6 molar fraction will cause the radial diffusion path of the UF_6 in the auxiliary gas to be prolonged because of the higher UF_6 speed ratio, which improves isotope separation (Sect.4.2.3, Fig.4.9). On the other hand, at a given Knudsen number, isotope separation will develop more quickly over the angle of deflection, if the UF_6 speed ratio is increased (Sect.4.2.3, Fig.4.8). Accordingly, for a given nozzle, any reduction in the UF_6 molar fraction requires the inlet pressure to be raised so that the transient maximum of the elementary effect of isotope separation is attained precisely at the point of flow splitting by the skimmer.

With decreasing UF_6 molar fraction, the ternary diffusion coefficient D_T converges against the binary diffusion coefficient D_m of the UF_6/auxiliary gas mixture. It follows from the ternary diffusion equations (4.17,18) that the relative difference between the partial cut θ_1 of the light isotope and the partial cut θ_u of the UF_6 and, hence, the elementary effect of isotope separation increases with the D_T/D_m ratio. This is due to the fact that reductions in the UF_6 molar fraction cause the number of UF_6-UF_6 collisions to become smaller and smaller compared with the number of UF_6/ auxiliary gas collisions, and the interaction between the two isotopic species to decline. The reduced interaction, in turn, favors the generation of different pressure-diffusion velocities of the isotopes in the auxiliary gas and, thus, the transient enhancement in the elementary effect of isotope separation. At high UF_6 molar fractions and correspondingly low values of the D_T/D_m ratio, the isotopic mixture as a whole strives to achieve its equilibrium distribution in the centrifugal field, since the relative difference in the radial diffusion velocities of $^{235}UF_6$ and $^{238}UF_6$ vanishes according to the higher the number of UF_6-UF_6 collisions. The transient enhancement in isotope separation will then be suppressed more and more strongly.

6.5 Operating Temperature

Both in technical separation nozzle systems and in the separation nozzles used for laboratory separation experiments and for probe measurements, the temperature of the nozzle walls is equal to the reservoir temperature T_0 of the process gas mixture. Consequently, T_0 is the relevant operating temperature of the separation process. Changing the operating temperature at given values of the inlet pressure and the expansion ratio will affect both the behavior of the flow and the diffusion processes in the separation nozzle, because of the changes in gas density and in the transport coefficients. Raising the operating temperature at a given inlet pressure results in an increase in the Knudsen number and in the diffusion coefficient, while the Reynolds number drops because of the decreasing mass throughput and the increasing viscosity. This immediately shows that the inlet pressure optimal for separation, p_0^+ , must rise along with T_0, so that the optimum values of the Knudsen number and the Reynolds number remain unchanged.

Figure 6.14 is a plot of a few typical results of separation experiments on H_2/UF_6 mixtures at various operating temperatures /72/. It is seen that raising the absolute temperature T_0 by some 20% results in a similar increase in the optimum operating pressure. Moreover, a very slight increase in the maximum elementary effect ε_A of isotope separation is seen with rising operating temperature.

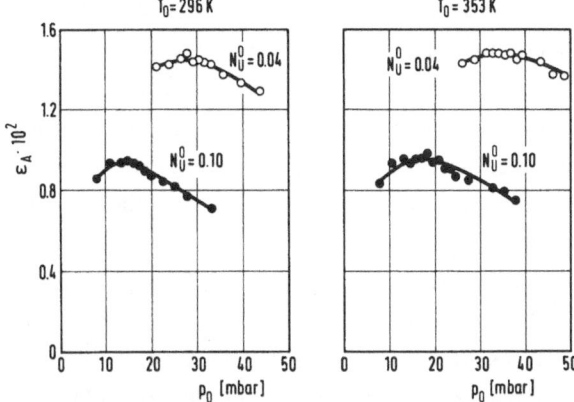

Fig.6.14. Elementary effect ε_A of isotope separation versus inlet pressure p_0 for various operating temperatures T_0 and UF$_6$ molar fractions N_u^0. Results of separation experiments using H$_2$/UF$_6$ mixtures in a standard separation nozzle; $\theta_u = 0.25$, $N_u^0 = 0.04$ and 0.10, $p_0/p' = 2.1$ at $N_u^0 = 0.04$ and $p_0/p' = 2.3$ at $N_u^0 = 0.10$, $p'' = p'$, $T_0 = 296$ K and 353 K, $r_0 = 0.75$ mm

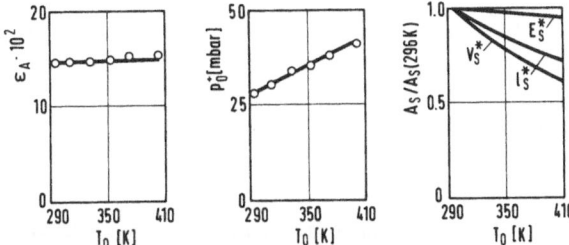

Fig.6.15. Influence of the operating temperature T_0 on the elementary effect $\varepsilon_A(p_0 = p_0^+)$ of isotope separation, the inlet pressure p_0^+ optimal for separation, and the specific process parameters A_s normalized to the values at 296 K. Results of separation experiments using an H$_2$/UF$_6$ mixture in a standard separation nozzle; $\theta_u = 0.25$, $N_u^0 = 0.04$, $p_0/p' = 2.1$, $p'' = p'$, $r_0 = 0.75$ mm

The experimental results shown in Fig.6.15 convey an idea of the influence of the operating temperature T_0 upon the elementary effect of isotope separation, $\varepsilon_A(p_0 = p_0^+)$, and upon the inlet pressure p_0^+ optimal for separation. The plot also shows the specific process parameters E_s^*, V_s^*, and l_s^*, normalized to the respective values at $T_0 = 296$ K; for the normalized specific energy consumption E_s^* a constant temperature independent of the operating temperature was assumed for calculating the isothermal compression work of the gas.[16]

---

[16] The operating temperature of the separation nozzles could be increased technically at a constant intake temperature of the compressor of a separation stage by feeding the hot gas from the compressor right into the separation nozzles and carrying off the heat of compression downstream from the separation nozzles. Unlike the separation stage concept shown in Fig.3.1, the gas cooler would then be located on the low-pressure side of the compressor.

In the temperature range between 300 and 400 K, $\varepsilon_A(p_0=p_0^+)$ rises by 3%-4%, while p_0^+ increases by 50%. In accordance with the increase in ε_A, the specific energy consumption based on constant compression work drops by some 6% - 8%. A much more pronounced temperature dependence is found for the specific suction volume, which drops by some 40% in the range of temperatures studied here because, in addition to the slight increase in ε_A, there is a marked increase in p_0^+ with T_0. The gas throughput per unit length of the nozzle slit rises with T_0 if the nozzle is operated at the inlet pressure optimal for separation; this results in a decrease of the specific slit length by almost 30% between T_0 = 300 and 400 K.

If the separation nozzle is operated at usual expansion ratios ($1.5 < p_0/p' < 3$) on H_2/UF_6 mixtures ($0.02 < N_u^0 < 0.1$) in the temperature range between 300 and 400 K, the following empirical relations can be derived from the separation experiments for the temperature dependence of separation properties and operating parameters /72/:

$$
\begin{aligned}
p_0^+(T_0) &\sim T_0^{1.3\pm0.3} & E_s(p_0=p_0^+) &\sim T_0^{-0.2\pm0.1} \\
\varepsilon_A(p_0=p_0^+) &\sim T_0^{0.1\pm0.05} & V_s(p_0=p_0^+) &\sim T_0^{-1.5\pm0.4} \\
L_m(p_0=p_0^+) &\sim T_0^{0.8\pm0.2} & l_s(p_0=p_0^+) &\sim T_0^{-1.0\pm0.3}
\end{aligned}
\tag{6.4}
$$

These relations are based on the assumption that the expansion ratio, the UF_6 molar fraction and the UF_6 cut are kept constant when the operating temperature T_0 is changed. It will be shown below that satisfactory explanations can be found for these empirical relations, if the temperature dependences of the sonic velocity, of the viscosity of the mixture, of the binary diffusion coefficient D_m and of the ternary diffusion coefficient D_T are taken into account /72/.

From (4.20,21) it follows that separation remains unchanged at a given speed ratio, if the quantity $D/(r_0 v_0)$, which is the reciprocal of the Péclet number of mass transfer, remains constant. This is tantamount to the fact that the ratio of the isotope transport normal to the direction of flow, which is proportional to vD, and the isotope transport in the direction of flow, which is proportional to vv_0r_0, is not changed, as was explained in Sect.5.1. In a gaseous mixture, the temperature dependence of the diffusion coefficient D at constant pressure can be described by a relation such as

$$
D = \text{const } T^{1+0.5+\delta} \quad .
\tag{6.5}
$$

The exponents of T characterize the effects of decreasing number density (T^1), increasing thermal velocity ($T^{0.5}$), and the reduction of the collision cross section for hard-sphere molecules (T^δ) associated with a temperature increase.

In an H_2/UF_6 mixture with a UF_6 molar fraction of 0.05, δ is around 0.3 in the range of temperatures between 300 and 400 K. If it is further assumed that the reference velocity v_0 changes with $\sqrt{T_0}$ like the thermal velocity, the temperature dependence of $D/(v_0 r_0)$ in this case turns out to be

$$D/(v_0 r_0) = \text{const } T_0^{1.3} \quad . \tag{6.6}$$

Since the diffusion coefficient changes inversely proportional to the number density, the effect of a change in T_0 on $D/(v_0 r_0)$ can be compensated by a change in the operating pressure of the separation nozzle; the inlet pressure optimal for separation, p_0^+, must be raised as

$$p_0^+ = \text{const } T_0^{1+\delta} \tag{6.7}$$

if T_0 is increased.

Another simplification can be made which assumes that the product ρD of the mass density and the diffusion coefficient has the same temperature dependence as the dynamic viscosity η, i.e., that the Schmidt number, $Sc = \eta/(\rho \cdot D)$, is independent of temperature. In this case, the Reynolds number of the flow,

$$Re = \frac{\rho v_0 r_0}{\eta} = \frac{\rho D}{\eta} \cdot \frac{v_0 r_0}{D} = \frac{1}{Sc} Pe \tag{6.8}$$

remains unchanged, if the Péclet number of mass transfer is kept constant. So if the optimum inlet pressure p_0^+ is increased with the operating temperature T_0 in accordance with (6.7), the speed ratio remains constant at a given expansion ratio, within the limits of the approximation made here. However, the optimum gas throughput, $L_m^+ = L_m(p_0 = p_0^+)$, clearly rises with increasing operating temperature, and it holds that

$$L_m^+ = \text{const} \cdot v^+(T_0)\, v_0(T_0)\, r_0 = \text{const}' \cdot T_0^{0.5+\delta} \quad . \tag{6.9}$$

The slight increase in the elementary effect of isotope separation with rising operating temperature cannot be explained by an approach which, for $p_0 = p_0^+(T_0)$, leads to constant values of the speed ratio and the Péclet number. Instead, to explain this effect, it is necessary to take into account the different potential parameters of the individual combinations of collision partners, which lead to different temperature dependences of the transport coefficients of a gaseous mixture. This shows that the D_T/D_m ratio of the ternary and the binary diffusion coefficients rises with temperature, because the binary diffusion coefficient $D_{1,h}$ of $^{235}UF_6$ and $^{238}UF_6$ increases more strongly with temperature than the diffusion coefficient $D_{a,u}$ of the mixture of auxiliary gas and UF_6 (4.15). The consequence is that as the operating temperature rises, the gas kinetic interaction between $^{235}UF_6$ and $^{238}UF_6$ is reduced relative to the interaction of the isotopic molecules with the auxiliary gas. This favors the development of different pressure diffusion velocities and thus the transient enhancement of the separation effect [Sect.6.4 and (4.17,18)].

In mixtures with low UF_6 molar fractions ($N_u < 0.05$), the viscosity η increases more weakly as a function of temperature than does the product ρD. As a result, the Schmidt number decreases with increasing temperature. According to (6.8) the Reynolds number must increase slightly with temperature if the Péclet number is kept constant. This in turn results in a light rise of the speed ratio and, hence, in a slight increase in isotope separation with rising operating temperature.

Within the framework of a more detailed analysis it can be shown that the small change in the ratio of the specific heats of the mixture between $T_0 = 300$ and 400 K and the resultant change in flow velocity can be neglected regarding the effects mentioned so far. Also the changes in the Prandtl number, which are around 5% in the temperature range between 300 and 400 K, are insignificant, because losses due to thermal conduction are only of secondary importance in the separation nozzle flow (Sect.5.2).

6.6 Type of Light Auxiliary Gas

The choice of the auxiliary gas in the separation nozzle process is determined primarily by the necessary conditions of low molecular weight and chemical compatibility with UF_6. In addition, aspects of process technology play an important role. For instance, if He is used instead of H_2, no explosive mixtures can be formed in case of air entering the separation stages. Moreover, a gaseous fluorinating agent can be added to a He/UF_6 mixture; this can prevent the formation of powdery solid reaction products from UF_6, such as UF_5 or UO_2F_2, which can most adversely affect the separation properties of the tiny nozzles.

The analysis of flow and separation experiments presented in Sects.4 and 5 indicates that the separation characteristics of various UF_6/auxiliary gas mixtures depend not only on the molecular weight of the auxiliary gas, but also on the potential of molecular interaction, which determines the transport coefficients of the mixture. The separation process is determined in particular by the diffusion coef-

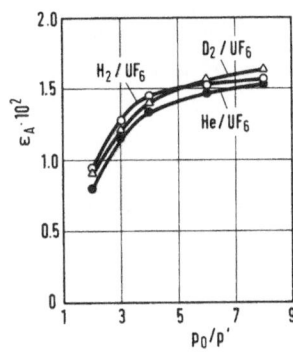 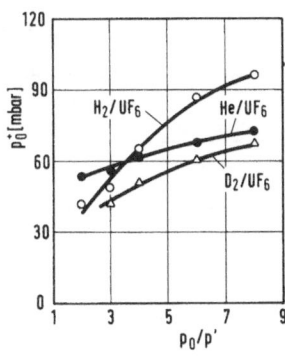

Fig.6.16. Elementary effect $\varepsilon_A(p_0=p_0^+)$ of isotope separation and optimum inlet pressure p_0^+ versus expansion ratio p_0/p' for H_2/UF_6, He/UF_6 and D_2/UF_6 mixtures. Results of separation experiments on a standard separation nozzle; $\theta_u = 0.33$, $N_u^0 = 0.05$, $p'' = p'$, $T_0 = 293$ K, $r_0 = 0.75$ mm, $a = 0.2$ mm

ficient of the mixture and by the ternary diffusion coefficient as well as by the viscosity of the mixture.

Figure 6.16 is a plot of a few typical results of separation experiments using H_2, He and D_2 as light auxiliary gases /73/; in all cases the UF_6 molar fraction was $N_u^0 = 0.05$.[17] It is seen that H_2 and D_2 allow higher values of $\varepsilon_A(p_0=p_0^+)$ to be attained at all expansion ratios than He. Up to expansion ratios $p_0/p' < 5$, the ε_A values of the H_2/UF_6 mixture are higher than those of the D_2/UF_6 mixture, while at high expansion ratios ($p_0/p' > 5$) higher uranium isotope separation is achieved with D_2. It is also seen that the relative difference in ε_A values of H_2/UF_6 and He/UF_6 mixtures clearly drops with increasing expansion ratio. However, the relative difference in ε_A values of D_2/UF_6 mixtures and He/UF_6 mixtures changes but little with the expansion ratio. The use of H_2 causes the optimum inlet pressure p_0^+ to rise in a clearly more pronounced way with the expansion ratio than do auxiliary gases of higher molecular weights. For high expansion ratios ($p_0/p' > 4$), the optimum inlet pressure of the H_2/UF_6 mixture clearly exceeds the corresponding p_0^+ values of the He/UF_6 and D_2/UF_6 mixtures.

The differences in the separation characteristics of the various UF_6/auxiliary gas mixtures become understandable if the influence of the mean molecular weight $\overline{M}$, the viscosity of the mixture η, the binary diffusion coefficient D_m, and the ternary diffusion coefficient D_T, upon flow and separation processes are considered, on the one hand, at low UF_6 speed ratios ($S_u < 3$) and, on the other hand, at high UF_6

---

[17] Unlike the separation experiments described before, a separation nozzle system was used in this case, which had a nozzle width 33% smaller ($a = 0.2$ instead of 0.3 mm) with the same radius of deflection ($r_0 = 0.75$ mm). The inlet pressure optimal for separation is higher for the smaller nozzle width, because the optimum values of the Knudsen number and Reynolds number are attained at a correspondingly higher gas density in the flow (cf. also Fig.5.15 and Sects.4.2.3 and 5.1).

speed ratios ($S_u > 4$) /73/. These data are listed below for H_2/UF_6, He/UF_6 and D_2/UF_6 mixtures with a UF_6 molar fraction of $N_u^0 = 0.05$ for a temperature of 293 K.

Mixture (N_u^0 = 0.05, T = 293 K)	H_2/UF_6	He/UF_6	D_2/UF_6
$\overline{M} = N_u^0 M_u + (1-N_u^0) M_a$	19.5	21.4	21.4
$\eta/\mu P$	150	240	168
$\nu D_m / \left(10^{18} \frac{particles}{cm\ s}\right)$	8.49	7.38	6.00
$\nu D_T / \left(10^{18} \frac{particles}{cm\ s}\right)$	4.22	3.94	3.53
D_T/D_m	0.50	0.53	0.59

At low UF_6 speed ratios, the elementary effect of isotope separation increases steeply with increasing S_u and, correspondingly, with increasing expansion ratio (see, e.g. Sect.4.2.3). Consequently, at a given expansion ratio and a given UF_6 molar fraction, the highest value of ε_A will be found here with the auxiliary gas resulting in the highest value of S_u. Since the mean molecular weight is lowest for the H_2/UF_6 mixture, it follows that the UF_6 speed ratio,

$$S_u = S_m \sqrt{M_u/\overline{M}} \qquad (6.10)$$

is about 5% higher than in the D_2/UF_6 and the He/UF_6 mixtures, if identical values of the speed ratio of the mixture S_m and a UF_6 molar fraction of $N_u^0 = 0.05$ are assumed. Moreover, in an H_2/UF_6 mixture, the dissipative losses due to viscous effects are smaller than in a He/UF_6 and a D_2/UF_6 mixture if, for all mixtures, comparable conditions are assumed for the separation process. This becomes clear if one looks at the relation already explained in Sect.6.5 between the Reynolds, Schmidt, and Péclet numbers of the mixture (6.8). It follows from the data compiled in the table that, at given values of the Péclet number and the speed ratio, i.e., at comparable conditions for separation, the Schmidt number is lowest for the H_2/UF_6 mixture and the Reynolds number is highest. Also the higher isotope separation in the D_2/UF_6 mixture, compared with the He/UF_6 mixture, at low expansion ratios is mainly due to the lower Schmidt number of the D_2/UF_6 mixture.

At high UF_6 speed ratios, the elementary effect of isotope separation increases only slightly with S_u and, consequently, only very little with the expansion ratio, thus causing the influences of viscous effects and mean molecular weight upon ε_A to lose significance more and more. The highest value of ε_A, at a given UF_6 molar fraction N_u^0, is found with the auxiliary gas at which the interaction among the UF_6 molecules relative to the interaction with the molecules of the auxiliary gas is a minimum. The favorable separation characteristics of the D_2/UF_6 mixture at high expansion ratios thus can be explained by the fact that, in this mixture, the ratio of D_T/D_m, which determines the difference in radial pressure diffusion velocities, is clearly higher than in the He/UF_6 and the H_2/UF_6 mixtures. It also becomes evident that the relative difference in the ε_A values of He/UF_6 and H_2/UF_6 mixtures becomes increasingly smaller with rising expansion ratio, because of the declining influence of S_u. At very high expansion ratios, because of the slightly higher D_T/D_m ratio of He, there may even be higher ε_A values with He than with H_2. The influence of D_T/D_m upon ε_A becomes increasingly smaller with decreasing UF_6 molar fraction, because D_T/D_m converges against 1 for all mixtures for infinitesimal values of N_u^0.

Considerations analogous to those applying to the elementary effect of isotope separation can also be used to explain the dependence of the optimum inlet pressure p_0^+ on the expansion ratio for the different auxiliary gases. The high value of p_0^+ resulting for He/UF_6 mixtures at low expansion ratios is due to the high viscosity of the mixture. Sufficiently high values of the Reynolds number and thus of the UF_6 speed ratio in the He/UF_6 mixture are achieved only at clearly higher gas densities than in the H_2/UF_6 mixture. At high expansion ratios it must be taken into account that the diffusion coefficient D_m and the UF_6 speed ratio S_u are higher in the case of H_2 than in the case of He, if the same number densities and UF_6 molar fractions are assumed. According to diffusion theory, the maximum of ε_A is attained in the H_2/UF_6 mixture at a smaller angle of deflection than in the He/UF_6 mixture, if equal values for the inlet pressure are applied. This, in turn, leads to the conclusion that at high expansion ratios the inlet pressure optimal for separation of the H_2/UF_6 mixture exceeds that of the He/UF_6 mixture. In the D_2/UF_6 mixture, both the viscosity and the νD_m product are clearly lower than in the He/UF_6 mixture, thus always leading to a lower value for p_0^+ in D_2 than in He.

In order to assess the various technological and economical aspects in respect to the alternative use of H_2/UF_6 and He/UF_6 mixtures in technical separation nozzle facilities, detailed measurements of the separation characteristics were carried out with these mixtures at various UF_6 molar fractions. In contrast to the measurements shown so far, a so-called biradial separation nozzle was used, in which the radius of curvature of the deflection wall was 0.5 mm in the first half of deflection and

0.75 mm in the second half (Sect.8.1).[18] In a detailed analysis of these measurements it was found that in H_2/UF_6 mixtures the specific energy consumption passes through a flat minimum in the molar fraction range between $N_u^0 = 0.05$ and $N_u^0 = 0.02$, while the specific slit length and the specific suction volume decrease clearly with decreasing UF_6 molar fraction in this range. In contrast, in He/UF_6 mixtures the specific energy consumption clearly increases in this range of molar fractions with decreasing N_u^0. Since the specific energy consumption is particularly important for the economics of the separation nozzle process, the advantages associated with a reduction in the other specific process parameters are overcompensated. The most advantageous UF_6 molar fraction with respect to overall economics is around 0.02 for the H_2/UF_6 mixture and around 0.04 for the He/UF_6 mixture.

The fact that there are different values of the optimum UF_6 molar fractions of H_2/UF_6 and He/UF_6 mixtures has several causes. With decreasing UF_6 molar fraction, the relative decrease in the mean molecular weight and, hence, the relative increase in the UF_6 speed ratio is more pronounced in case of the H_2/UF_6 mixture. Moreover, the viscosity decreases relatively more strongly in the transition to low N_u^0 values in the H_2/UF_6 mixture than in the He/UF_6 mixture; at $N_u^0 = 0.05$, the viscosities of H_2/UF_6 and He/UF_6 mixtures behave as 1:1.6, and at $N_u^0 = 0.02$ as 1:1.75. Finally, the relative increase in D_T/D_m associated with a reduction of N_u^0 is also slightly more pronounced in the H_2/UF_6 mixture than in the He/UF_6 mixture.

The most important results of these comparative studies performed on a biradial separation nozzle are shown in Fig.6.17. The curves represent the elementary effect ε_A of isotope separation and the specific process parameters, E_s^{id}, V_s^{id} and l_s^{id}, as a function of the expansion ratio p_0/p' at the inlet pressure optimal for separation, p_0^+. For the reasons of economy mentioned above, the UF_6 molar fraction was set at $N_u^0 = 0.02$ for the H_2/UF_6 mixture and at $N_u^0 = 0.042$ for the He/UF_6 mixture.

The minimum value of specific energy consumption of the H_2/UF_6 mixture, which is attained at relatively low expansion ratios because of the low UF_6 molar fraction, is about 20% below that of the He/UF_6 mixture. At an expansion ratio of $p_0/p' = 3$, the specific energy consumption of both UF_6/auxiliary gas mixtures is about the same, but the specific suction volume and the specific slit length of the H_2/UF_6 mixture are, respectively, 50% and 70% below the corresponding values of the He/UF_6 mixture. The number of stages of a separation nozzle cascade, which is inversely

---

[18] The biradial separation nozzle has clearly more advantageous separating characteristics than the standard separation nozzle used in earlier measurements. The more recent measurements thus update the comparison of separation characteristics of He/UF_6 and H_2/UF_6 mixtures, taking into account the advancement in the separation nozzle process that has taken place in the meantime.

proportional to ε_A, is about 30% smaller for the H_2/UF_6 mixture than for the He/UF_6 mixture if $p_0/p' = 3$ is assumed for both mixtures.

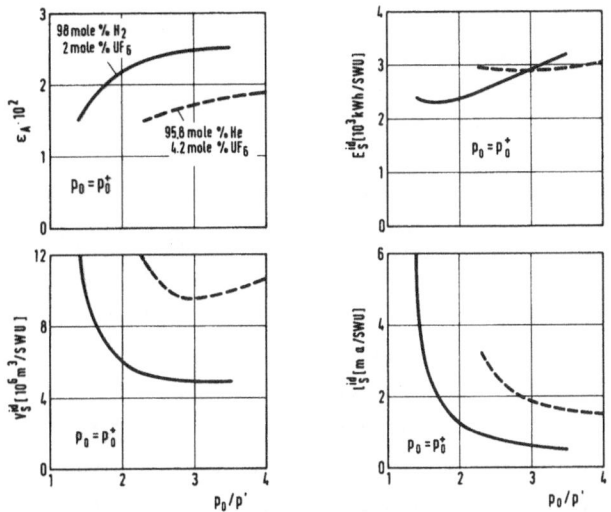

Fig.6.17. Comparison of the separation characteristics of H_2/UF_6 and He/UF_6 mixtures at economically favorable UF_6 molar fraction N_u^0 for the respective auxiliary gas, representing the elementary effect $\varepsilon_A(p_0 = p_0^+)$ of isotope separation, the specific energy consumption $E_s^{id}(p_0=p_0^+)$, the specific suction volume $V_s^{id}(p_0 = p_0^+)$, and the specific slit length $l_s^{id}(p_0 = p_0^+)$ versus the expansion ratio p_0/p'. Results of separation experiments on a biradial separation nozzle; $\theta_u = 0.25$, $N_u^0 = 0.02$ for H_2/UF_6 and $N_u^0 = 0.042$ for He/UF_6, $p'' = p'$, $T_0 = 296$ K, $r_0 = 0.5$ mm/0.75 mm, for the nozzle contour see Fig.8.1

7. Measuring Technique and Experimental Strategy Applied in Optimizing Separation Nozzle Systems

Since the theoretical methods available at present do not permit a quantitative evaluation to be made of the coupled flow and diffusion processes in a separation nozzle, the operating conditions and geometric parameters of separation nozzle systems can be optimized only on the basis of systematic separation experiments using UF_6/auxiliary gas mixtures. This results in an extremely lengthy optimization program for each individual separation nozzle geometry, because the different operating conditions cannot be optimized separately. For instance, if one considers one set of operating conditions resulting in a minimum of the specific energy consumption E_s at a given UF_6 molar fraction N_u^0, it directly follows from the studies outlined in Sect.6 that changing to a different value of N_u^0 will also change the values of the inlet pressure, the expansion ratio and the stagnation ratio pertaining to the new minimum of E_s. Consequently, the operating pressures assigned to the new minimum of E_s can be determined only by a relatively large number of separation experiments. When optimizing the operating conditions of a technical separation nozzle plant, the different contributions by the individual specific process parameters E_s, V_s, l_s to the overall technical expenditure for uranium enrichment must be taken into account (Sect.3). Since the specific process parameters change differently with the operating conditions, such a problem can be solved only by multidimensional optimization, in which all operating conditions are varied over wide ranges.

Carrying out such optimization programs for various types of separation nozzles (Sect.8) at a tolerable expenditure, on the one hand requires minimizing the measuring time needed per separation experiment. On the other hand, high measuring accuracy is a prerequisite to detecting reliably potential improvements, i.e., when optimizing a geometric parameter. Finally, in order to limit the number of experiments, it is important to perform separation experiments only at those values of the UF_6 cut which could be used in a separation nozzle cascade (Sects.3 and 6.3).

Figure 7.1 represents a highly simplified flowsheet of equipment for optimization experiments on laboratory-type separation nozzles with UF_6/auxiliary gas mixtures /74,75/. The mode of operation of this equipment, which largely meets the re-

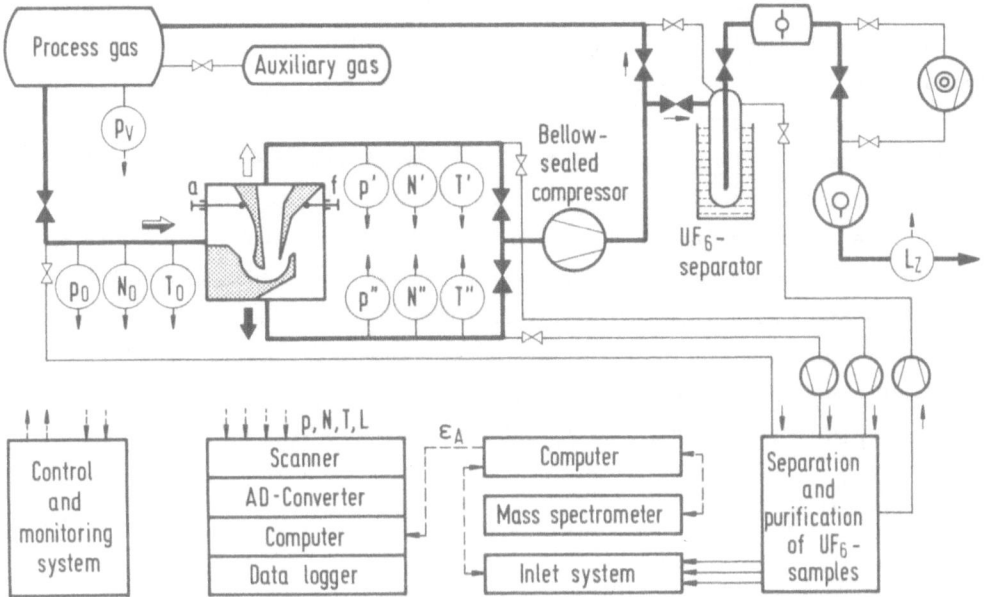

Fig.7.1. Schematic representation of equipment for separation experiments on labora-tory-scale separation nozzles with UF_6/auxiliary gas mixtures

quirements listes above with respect to optimizing the operating conditions of sep-aration nozzle systems, will be briefly described below.

In the separation experiments, the process-gas mixture, which is stored in sev-eral tanks at different UF_6 molar fractions at a pressure of p_V, is throttled to the nozzle inlet pressure p_0 by means of a control valve, fed to the separation nozzle and split into a light and a heavy fraction. Downstream from the control valves used to set the suction pressures p' and p" of the light and heavy fractions, the two par-tial streams are recombined. The gas mixture is precompressed in a compressor resist-ant to UF_6, the UF_6 is separated from the auxiliary gas in cryotraps, and the auxil-iary gas is pumped to the outside by means of a rotary vane pump. The throughput L_z of the auxiliary gas through the separation nozzle is determined in the exhaust pipe by means of a thermal flowmeter and, in a redundant fashion, by means of a volumetric gas meter. In an alternative operating mode, the process gas can be pumped back in-to the storage tank without the UF_6 being separated from the auxiliary gas.

The separation nozzle systems used for the laboratory scale separation experiments are mostly designed in such a way that two geometric parameters, e.g., the nozzle width a and the skimmer distance f, can be changed without opening the separation element chamber. The adjustment of the skimmer distance is of particular importance in optimization, since it allows the UF_6 cut to be set to a given value.

In the gas mixture fed to the separation nozzle and in the gas streams of the light and the heavy fractions, the UF_6 molar fractions N_u^0, N_u' and N_u'', are measured continuously in addition to the pressures and the temperatures. The composition of the different gas streams is determined by two independent measuring methods based on the dependence of the average ionization cross section on the gas composition and on the selective absorption of infrared radiation by UF_6 at a wavelength of 16 μm /76/. The measuring systems developed for analyzing the mixtures of auxiliary gases and chemically most aggressive UF_6 have a relative error of approximately ± 1% in the range of interest for optimization. Redundant analysis of the mixture on the basis of two different measuring principles allows systematic errors in determining the UF_6 molar fraction to be recognized quickly.

From the UF_6 molar fractions, the UF_6 cut,

$$\theta_u = \frac{N_u'(N_u''-N_u^0)}{N_u^0(N_u''-N_u')} \quad , \tag{7.1}$$

and the separation factor of the mixture,

$$A_m = \frac{N_u''(1-N_u')}{N_u'(1-N_u'')} \tag{7.2}$$

are calculated and indicated continuously. In this way it is possible to set at a given value the UF_6 cut by changing the skimmer position or changing the operating pressures, thus avoiding separation experiments at UF_6 cuts not eligible for separation nozzle plants. Moreover, direct observation of the separation of UF_6 and auxiliary gas already furnishes important information about the properties of the separation system; as a consequence, isotopic analysis, which takes much time, can partly be avoided.

After the desired operating conditions have been set, process gas samples are automatically taken from the feed gas and the light and the heavy fractions for isotopic analysis in a mass spectrometer. For this purpose, the UF_6 is first separated from the auxiliary gas in small sampling reservoirs cooled with liquid nitrogen. Next, the UF_6 is gradually heated and gaseous impurities, especially the hydrogen fluoride always present in the process gas, are removed by pumping until a temperature of about 230 K has been reached. The UF_6 samples processed and purified in this way are heated to room temperature and fed into a computer-controlled magnetic mass spectrometer, by which the isotopic ratios I_0, I' and I'' in the three samples are determined. The UF_6 not consumed for isotopic analysis is pumped into the UF_6 separator, and the sampling reservoirs are heated to about 400 K during pumping in order to avoid memory effects. The whole sampling system consists of three independent units, i.e., three sampling reservoirs are operated in parallel for cryo-separation and purification of UF_6, three for isotopic analysis, and three for evacuation, thus ensuring optimum utilization of the mass spectrometer.

Measurements of the relative difference in the isotopic ratios of two UF_6 samples by mass spectrometry are usually associated with an error of about $\pm 1 \times 10^{-4}$. Accordingly, the values measured for the elementary effect of isotope separation,

$$\varepsilon_A = \frac{n'(1-n'')}{n''(1-n')} - 1 = \frac{I'}{I''} - 1 \quad , \tag{7.3}$$

have a relative error of about $\pm 1\%$, if ε_A values of 0.01 to 0.02, which are typical of the separation nozzle process, are used as a basis (n' and n'' are the $^{235}UF_6$ molar fractions and I' and I'' are the isotopic ratios in the UF_6 of the light and the heavy fractions, respectively). Measuring the isotopic ratios I_0, I' and I'' allows the UF_6 cut to be determined by a relation analogous to (7.1); this permits the θ_u values determined from the UF_6 molar fractions N_u^0, N_u' and N_u'' to be verified by a redundant measurement.

In order to keep the average time spent on a separation experiment as short as possible, the experimental procedure must be adapted to the specific boundary conditions of the separation equipment in such a way that besides the time required for measurements, the time spent on setting certain operating parameters is also minimized. Since setting the operating pressures takes the shortest time, first the nozzle inlet pressure p_0 and the expansion ratio p_0/p' are varied at a given skimmer position in such a way that the UF_6 cut θ_u is set to the desired value.[19] Such measurements are performed at different skimmer positions until a sufficient number of data exist which allow one to determine accurately the relations $\varepsilon_A(p_0, p_0/p')$ and $L_u(p_0, p_0/p')$ and, hence, the corresponding relations for the specific process parameters within the ranges of interest for the respective optimization program; this procedure has been illustrated by the contour line plots shown in Figs.6.2 and 6.3. After such a cycle of measurements, another operating parameter is changed, e.g., the nozzle width a or the UF_6 molar fraction N_u^0, and again the nozzle inlet pressure and the expansion ratio are varied.

In performing such an optimization program it is most important that the specific process parameters be determined without delay for each test setting by means of the computer system, which is part of the equipment. This allows information to be ob-

---

[19] The curves shown in Fig.6.4 exemplify such variations of the inlet pressure and the expansion ratio at various skimmer distances f for a constant UF_6 cut of $\theta_u = 0.25$.

tained about the dependence on operating conditions of the separative work costs of an industrial facility, if corresponding weighting factors are assigned to the different specific process parameters. Such analysis of the experiments on the basis of technical and economic aspects ensures that the range of variations of the different geometric and operating parameters of a separation nozzle system is limited appropriately and allows the number of separation experiments to be restricted in a reasonable way.

8. Separation Nozzle Designs

In the previous sections, the diffusion and flow processes in the separation nozzle were represented essentially by the example of the standard separation nozzle for two reasons. On the one hand, the standard separation nozzle for a long time constituted the basis of technical implementation of the separation nozzle process and, for this reason, is the system most thoroughly studied. On the other hand, it can be used as a means of describing, in a relatively simple and clear way, the characteristic physics phenomena of the separation nozzle method.

In Sect.2 it has been pointed out that there are various other separation nozzle designs besides the standard separation nozzle. The systems found to be most successful in the development of the separation nozzle process will be outlined below.

In Sect.8.1, a system will first be described in which the standard principle of the "mechanical" deflection of the flow by a curved wall has been retained, but which is characterized by a multitude of geometric changes of the nozzle contour. These changes have resulted in particular in a considerable reduction of specific energy consumption compared with the standard separation nozzle.

In Sect.8.2, the so-called double-deflection system is described in which the heavy fraction of a first nozzle with single mechanical flow deflection is fed right into a second nozzle for further separation.

This additional separation does not require any extra expenditure in terms of compression work, because the high stagnation pressure of the heavy fraction is utilized, which considerably exceeds the suction pressure of the light fraction and, hence, the intake pressure of the compressor of the separation stage (Sect.6.2). The double-deflection system can especially help to reduce considerably the number of separation stages required in an enrichment plant.

Besides the separation nozzle systems considered so far, in which the separating centrifugal field is generated by deflecting the flow at a fixed curved wall, separation nozzle systems with dynamic flow deflection have been developed. In these arrangements, which are described in Sect.8.3, the centrifugal forces are not acting upon a fixed wall, but use is made of the balance of forces in curved flows arranged in mirror symmetry relative to each other.

8.1 Systems with Single Mechanical Flow Deflection

The main characteristic of separation nozzles with single mechanical deflection is
seen in the fact that a two-dimensional flow is deflected at a concave curved wall
and subsequently separated into two fractions by means of a wedge-shaped flow split-
ter, the so-called skimmer. The angle of deflection of the flow is mostly around
180°; in the first part of deflection, the flow is generally confined by a curved
duct formed by the deflection wall and the inner nozzle wall. A particularly typical
version of such a system, as has been emphasized repeatedly above, is the standard
separation nozzle shown in Fig.2.1.

The nozzle contour of the standard separation nozzle has mainly been developed
empirically in separation tests using mixtures of UF_6 and auxiliary gas. Extensive
fundamental studies have indicated many approaches for further advancements of the
concept based on single mechanical flow deflection. A combination of such improvements has been employed in the system compared with the standard separation nozzle
in Fig.8.1. Its geometry differs in the following respects:

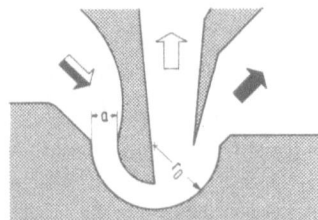 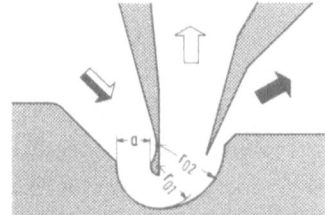

Fig.8.1. Contours of a standard separation nozzle (a = 0.3 mm, r_0 = 0.75 mm) and
a biradial separation nozzle (a = 0.4 mm, r_{01} = 0.5 mm, r_{02} = 0.75 mm) used for
separation experiments with auxiliary gas/UF_6 mixtures

a) The radius of curvature of the inner nozzle wall of the improved system is much
smaller than that of a standard separation nozzle having the same average radius of
curvature of the deflection wall.
b) The radius of curvature of the deflection wall is enlarged in the second half of
the deflection process.
c) The skimmer is inclined more strongly relative to the direction of the upstream
flow than in the standard separation nozzle.
d) The inner wall of the nozzle has a pronounced groove in the inlet section.

Since the more recent system has two different radii of curvature of the deflection
wall as a particular characteristic, it will be called the biradial system below.

The considerations leading to these modifications of the nozzle contour and the basic mechanisms underlying the corresponding improvements in the separation characteristics will be explained in the following section. All considerations are based on the criteria that the radial diffusion paths of the UF_6 molar stream surfaces in the auxiliary gas should be as large as possible, and pressure losses in the inlet section during deflection and in the discharge duct of the light fraction should be as low as possible.

8.1.1 Criteria for Optimizing Nozzle Geometries

From the studies described in Sect.4.2.4 concerning the influence of the structure of the centrifugal field upon isotope separation, it first of all follows that the starting conditions for the separation process become more favorable if the UF_6 molar stream surfaces at the beginning of deflection are further removed from their equilibrium positions. Accordingly, for purely geometric reasons, the ratio of the radius of curvature r_1 of the inner nozzle wall to the radius of curvature r_0 of the deflection wall should be as small as possible, because reducing the r_1/r_0 ratio causes the UF_6 molar stream surfaces to be shifted towards the center of curvature at the beginning of deflection and, correspondingly, to be more distant from their equilibrium positions in the centrifugal field. Moreover, a small radius of curvature r_1 of the inner nozzle wall will reduce the viscous drag at the inner boundary of the curved flow and thus promote the generation of a flow profile in which the flux rises with decreasing radius, very much like a potential vortex.[20] This effect additionally supports the accumulation of UF_6 in the range of small radii of curvature desired at the beginning of deflection.

The radial diffusion path of a UF_6 molar stream surface in the auxiliary gas increases with the intensity of the centrifugal field upstream from the skimmer. According to (5.15), the intensity of the centrifugal field can be characterized by the difference Δp in static pressures at the outer and inner boundaries of the flow. Hence, the need to have a radial diffusion path as long as possible at a given Reynolds number directly results in the criterion of Δp being made as large as possible in the region directly upstream from the skimmer. In order to achieve high Δp values

---

[20] Since decreasing the ratio of r_1/r_0 reduces the effective length of the inner nozzle wall, the gas is accelerated at the inner boundary of the centrifugal field over a shorter distance; this has a similar effect on the flow profile as an increase in the Reynolds number in the case of a slender curved channel (Sect. 5.2 and Fig.5.1). This finding is also confirmed by free molecular probe measurements showing that decreasing r_1/r_0 causes the flux maximum to be shifted from the center of the inlet channel towards smaller radii.

it is favorable, on the one hand, if the flow is decelerated at the periphery of the centrifugal field, i.e., if the dynamic pressure of the outer flow regions is converted into static pressure. On the other hand, the static pressure should be as low as possible at the inner boundary of the centrifugal field, i.e., the light fraction should be extracted at a minimum pressure loss.

The stagnation of the outer flow regions desired for the separation process can be achieved, e.g., as explained in Sect.6.2, by raising the suction pressure p" of the heavy fraction to a level clearly above the suction pressure p' of the light fraction. A qualitatively similar change in the flow field is possible by inclining the skimmer so that the skimmer and the deflection wall constitute a channel with decreasing cross section in the direction of the flow. The cross section of the discharge duct for the light fraction will increase by a correspondingly greater margin in the flow direction. Such increase in the cross section causes the pressure losses in the light fraction to drop, unless there is a separation of the flow at the skimmer edge for large angles of attack of the skimmer; the vortices occurring in this case then again increase the pressure losses in the light fraction, as indicated by free molecular probe measurements. In order to obtain a low static pressure at the inner boundary of the curved flow, the cross section of the duct for the light fraction must be sufficiently large at the end of deflection. This requirement can be met by making the radius of curvature of the deflection wall larger in the second than in the first part of deflection, thus increasing the distance between the edge of the skimmer and the inner wall of the nozzle.

Besides the boundary conditions mentioned above, the angle of deflection is essential with respect to the pressure losses and the radial diffusion path of the UF_6 molar stream surfaces in the curved flow. Since the Reynolds number optimal for isotope separation rises roughly in proportion to the angle of deflection, and the relative pressure losses in the intake and discharge ducts of the nozzle decrease with increasing Reynolds number, the angle of deflection should be made large enough so that sufficiently wide cross sections are retained for the intake and discharge ducts. Consequently, the need to have an angle of deflection as large as possible means that the pressure losses in the intake and discharge ducts should be as small as possible relative to the total pressure losses in the separation nozzle. Furthermore, at the beginning of deflection, the maximum of the UF_6 flux is shifted towards smaller radii with increasing Reynolds number. Thus, an increase in the angle of deflection results in more advantageous starting conditions with respect to the radial diffusion paths of the UF_6 molar stream surfaces.

It follows from the studies described in Sect.4.2.2 that the radial diffusion path of the UF_6 molar stream surfaces increases with decreasing UF_6 cut, because the inner stream surfaces are further away from their equilibrium positions at the beginning of deflection than the outer stream surfaces. Consequently, the maximum of isotope separation is reached in the UF_6 stream surfaces close to the deflection wall at smaller angles of deflection than in the inner stream surfaces, despite the radial increase in density and the corresponding radial decrease of the diffusion coefficient. This fact has been demonstrated above by the results of molecular probe measurements shown in Fig.5.8, in which the separation of sulfur isotopes for a mixture of He/SF_6 was determined in the standard separation nozzle. In that specific example it was found that the elementary effect ε_A of isotope separation at the deflection wall hardly rises any more for angles of deflection in excess of 120°, while a clear increase in ε_A with the angle of deflection is observed at low cuts.[21] Accordingly, the criterion of having an angle of deflection of the flow in the separation nozzle as large as possible means in particular to make the angle of deflection of the inner stream surfaces as large as possible. Since the angle of deflection of these stream surfaces is mainly determined by the shape of the inner nozzle wall, it is obviously advantageous to retain the convex curvature of the inner nozzle wall also upstream of the angle of deflection $\phi = 0$ of the deflection wall as long as possible, i.e., to provide a groove in the inner nozzle wall in the inlet section.

As a matter of fact, the possibilities of improvement discussed here cannot simply be translated into quantitative modifications of certain geometric parameters. Thus, e.g., a change in the Reynolds number of the separation nozzle flow will generally entail a change in the skimmer position pertaining to a certain UF_6 cut. In this way, it affects both the stagnation of the flow at the deflection wall and the suction conditions of the light fraction, so that changes in the inlet pressure or in the expansion ratio are bound to entail also changes in the geometric parameters "optimal" for separation in a specific case. This makes optimizing a separation nozzle geometry a multidimensional optimization problem affected not only by geometric parameters, but also by all operating conditions.

---

[21] In this connection, it should be pointed out that the reverse separation brought about by the opposed curvature of the streamlines at the inlet edge of the deflection wall is not likely to have a particularly negative effect. In the outer stream surfaces, where this effect is particularly pronounced, the transient process of isotope separation proceeds more quickly because of the shorter diffusion path, so that reducing the effective angle of deflection by reverse separation has less significant consequences.

8.1.2 Influence of the Nozzle Geometry upon the Separation Characteristics and the Specific Expenditure

Figure 8.2 is a plot of the inlet pressure p_0^+ optimal for isotope separation and the elementary effect $\varepsilon_A(p_0=p_0^+)$ of isotope separation versus the expansion ratio p_0/p' of the light fraction for the separation nozzle systems shown in Fig.8.1. It is seen that the ε_A values of the biradial system are clearly above those of the standard separation nozzle. Because of the roughly 25% smaller nozzle width a of the standard separation nozzle, its optimum inlet pressure p_0^+ is slightly higher than that of the biradial system. However, the Reynolds number optimal for separation, which is roughly proportional to $p_0^+ \cdot a$, is slightly higher in the biradial system than in the standard separation nozzle, if the same expansion ratios are used as a basis for both nozzles.

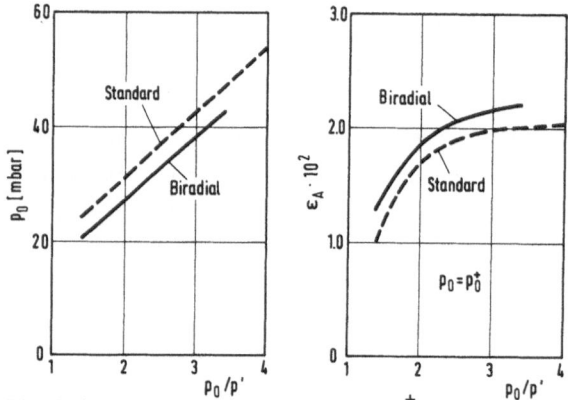

Fig.8.2. Optimum inlet pressure p_0^+ and elementary effect $\varepsilon_A(p_0 = p_0^+)$ of isotope separation versus expansion ratio p_0/p' for a standard separation nozzle and a biradial separation nozzle. Results of separation experiments using an H_2/UF_6 mixture; $\theta_u = 0.25$, $N_u^0 = 0.03$, $p'' = p'$, $T_0 = 296$ K; for the dimensions of the separation nozzles see Fig.8.1

Figure 8.3 is a plot of the specific process parameters of the separation nozzle systems shown in Fig.8.1 versus the expansion ratio p_0/p' of the light fraction, the inlet pressure always being optimal for separation. Comparing the two separation nozzle systems at the expansion ratio of $p_0/p' = 2$, which is typical for practical applications of the separation nozzle process, shows that the specific energy consumption of the biradial system is some 20%, the specific slit length some 25% below the corresponding values of the standard separation nozzle. The specific suction volume V_s^f for a fixed skimmer distance, whose significance in the fabrication of technical-scale separation nozzle elements was explained in more detail in Sect.6.1, is about 20% lower in the standard separation nozzle than in the biradial system.

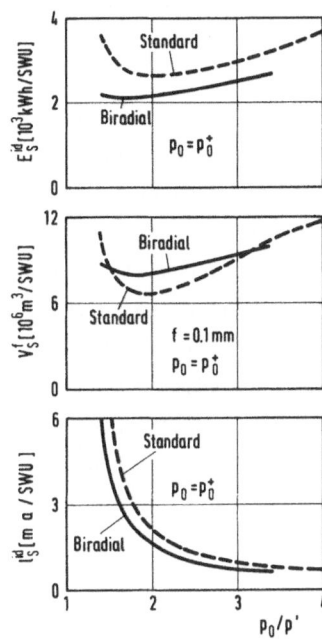

Fig.8.3. Specific energy consumption E_s^{id}, specific suction volume V_s^f, for a constant skimmer distance of 0.1 mm, and specific slit length l_s^{id}, versus expansion ratio p_0/p' at optimum inlet pressure p_0^+, for a standard separation nozzle and a biradial separation nozzle. Results of separation experiments using H_2/UF_6 mixture; for the experimental conditions, see Fig.8.2

The reduction of E_s^{id} in the biradial system is directly due to the higher separation effect ε_A, while the reduction of l_s^{id} is due both to the increase in the separation effect and the increase in the optimal Reynolds number which results in a correspondingly higher UF_6 throughput. The undesirable increase in V_s^f is mainly due to the fact that in the biradial separation nozzle system, not only isotope separation, but also the separation of UF_6 and the auxiliary gas is improved, i.e., the UF_6 is more strongly concentrated at the deflection wall. As a consequence, also the skimmer distance f required to set a particular UF_6 cut is smaller in the biradial system than in the standard separation nozzle, if the other dimensions of the system are fixed in such a way that the same values result for the optimum inlet pressure at a given expansion ratio. In this case, the volume flow for both systems is practically identical, while the separative work output of the biradial system is clearly higher, because of the higher elementary effect of isotope separation. In the specific suction volume V_s^f, obviously the advantage of the higher separative work output can be overcompensated by the disadvantage of the smaller skimmer distance.

On the basis of detailed economic analyses it can be shown that a change from the standard separation nozzle to the biradial system can greatly reduce the overall specific expenditure associated with the process and thus allow the separative work cost to be reduced by about 20%. Such economic analyses are based on detailed measurements of the separation characteristics on the one hand, and on the other hand, on planning work for industrial-scale facilities, from which can be determined the contributions the specific process parameters make to the operating costs and the capital costs of plants.

8.2 Systems with Double Mechanical Flow Deflection

8.2.1 Design and Mode of Operation

In the systems with single mechanical flow deflection, the mean stagnation pressure of the gas flow of the heavy fraction right after splitting of the flow by the skimmer is roughly identical to the inlet pressure p_0. This is due to the fact that the UF_6 accelerated by the auxiliary gas is concentrated at the deflection wall, which compensates the pressure losses due to dissipative effects by the increase in the mean molecular weight and the Mach number of the mixture (Sect.5.4). The difference between the mean stagnation pressure of the heavy fraction and the suction pressure of the compressor which is fixed by the pressure level of the light fraction can be utilized for further separation of the heavy fraction in a second separation nozzle. In this way, the total separative work production can be increased without additional expenditures in compression work /29,68,69/.

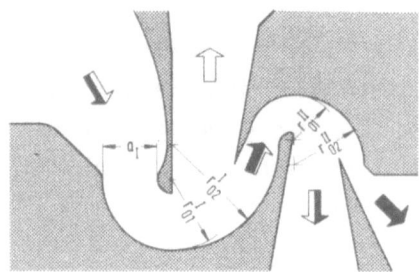

Fig.8.4. Nozzle contour of a separation nozzle system with double mechanical flow deflection (double-deflection system). Dimensions of the system used for the separation experiments: $a_I = 0.4$ mm, $r_{01}^I = 0.5$ mm, $r_{02}^I = 0.75$ mm, $r_{01}^{II} = 0.31$ mm, $r_{02}^{II} = 0.47$ mm

The nozzle contour of such a separation system with twofold mechnical flow deflection and trifractionation of the flow, which will be called the double-deflection system for short, is shown in Fig.8.4. In order to minimize viscous losses the gas flow of the heavy fraction of the first nozzle is fed into the second nozzle right after the flow has been split by the skimmer. Accordingly, the curvature of the streamlines in the second nozzle is opposite to that in the first nozzle. The width of the second nozzle, given by the position of the skimmer of the first nozzle, is clearly smaller than that of the first nozzle. Since the ratio of the nozzle width to the radius of curvature of the deflection wall should be as high as possible for efficient separation (Sect.8.1.1) the average radius of curvature of the deflection wall of the second nozzle is clearly smaller than that of the first nozzle. The contours of the two nozzles connected in series roughly correspond to those of the bi-radial separation nozzle system shown in Fig.8.1.

For use of the double-deflection system in a separation nozzle cascade, the operating conditions are best set in such a way that the isotopic ratio in the UF_6 strea of the light fraction of the second nozzle is identical to the isotopic ratio of the UF_6 feed stream into the separation stage. This partial stream of the double-deflection system, which will henceforth be called the intermediate fraction, can then be recycled within the separation stage without there being mixing losses. This mode of operation of the double-deflection system is shown in the separation stage principle sketched in Fig.8.5. If L_u^V is the UF_6 throughput through the compressor, the net UF_6 throughput through the separation stage is described by

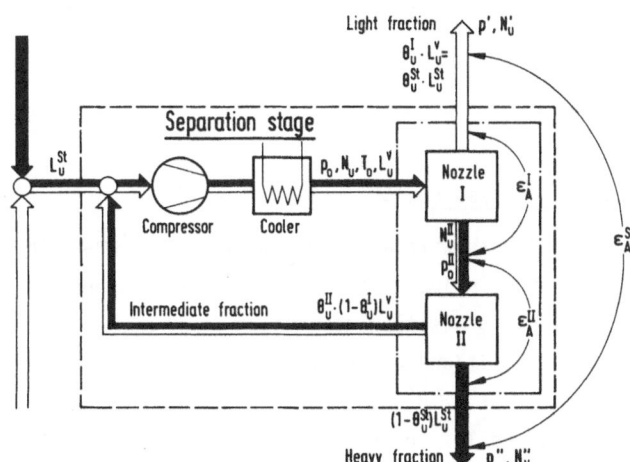

Fig.8.5. Flowsheet of a separation stage with internal reflux of the intermediate fraction of a double-deflection system

$$L_u^{St} = L_u^V \left[1 - (1-\theta_u^I)\theta_u^{II} \right] \quad . \tag{8.1}$$

The UF_6 cuts θ_u^I and θ_u^{II} as usual characterize the fraction of the UF_6 feed stream of the respective nozzle, which is discharged as the light fraction, i.e., θ_u^I is formed with the UF_6 throughput, L_u^V, of the compressor and θ_u^{II} is formed with the UF_6 feed stream, $(1-\theta_u^I)L_u^V$, of the second nozzle. The "stage UF_6 cut" referred to the net UF_6 throughput L_u^{St} of the separation stage,

$$
\begin{aligned}
\theta_u^{St} &= \theta_u^I \, L_u^V / L_u^{St} \\
&= \theta_u^I \left[1 - (1-\theta_u^I)\theta_u^{II} \right]^{-1} \quad ,
\end{aligned}
\tag{8.2}
$$

obviously will always be higher than the UF_6 cut θ_u^I of the first nozzle. The elementary effect of isotope separation of the stage,

$$\varepsilon_A^{St} = \frac{\theta_1^{St} - \theta_h^{St}}{\theta_h^{St}(1 - \theta_1^{St})} \tag{8.3}$$

is formed with the partial stage cuts θ_1^{St} and θ_h^{St} of the light and the heavy uranium isotopes, in analogy with (3.1). Neglecting higher order terms, it holds that

$$\varepsilon_A^{St} = \varepsilon_A^{I} + \theta_u^{II} \cdot \varepsilon_A^{II} \quad . \tag{8.4}$$

In this case, ε_A^{I} and ε_A^{II} are the elementary effects of isotope separation of the first and second nozzles. The separative work output of the stage, corresponding to (3.4), results directly from the stage separation effect ε_A^{St}, the net UF$_6$ through-put L_u^{St}, and the stage UF$_6$ cut θ_u^{St}:

$$\delta U^{St} = \frac{1}{2} \theta_u^{St}(1 - \theta_u^{St}) L_u^{St}(\varepsilon_A^{St})^2 \quad . \tag{8.5}$$

8.2.2 Influence of Operating Conditions on Isotope Separation

Some typical results of separation experiments with an H$_2$/UF$_6$ mixture are shown in Figs.8.6,7. The measurements illustrate the influence of the inlet pressure and the expansion ratio upon the elementary effects of isotope separation in the first and second nozzles and the whole system. The positions of the skimmers in the two nozzles were always set in such a way that a cut of $\theta_u^{I} = 0.235$ resulted for the first nozzle and of $\theta_u^{II} = 0.386$ for the second nozzle. On the one hand, this ensures that the first nozzle is operated roughly at the UF$_6$ cut optimal in terms of separative work output (Sect.6.3, Fig.6.10). On the other hand, a cut of $\theta_u^{St} = 1/3$ results for the separation stage in accordance with (8.2), which means that the requirements of a simple cascade arrangement are taken into account (Sect.3).

While the dependence of the separation effect ε_A^{I} of the first nozzle on the inlet pressure and on the expansion ratio reflects the behavior known from the standard separation nozzle (Sect.6.1, Figs.4.11, 6.1,2), the separation effect ε_A^{II} of the second nozzle shows a fundamentally different behavior. One particular characteristic is that ε_A^{II} passes through a minimum around the range of inlet pressures in which ε_A^{I} reaches its highest value and that ε_A^{II}, unlike ε_A^{I}, decreases with rising expansion ratio. The absolute values of ε_A^{II} are about 2 to 5 times smaller than those of ε_A^{I} at the operating conditions given in this case. The behavior of the separation effect ε_A^{St} of the double-deflection separation stage therefore qualitatively corresponds to the behavior of ε_A^{I}.

The separation behavior of the second nozzle of the double-deflection system can be explained plausibly by looking more closely at the starting conditions for the

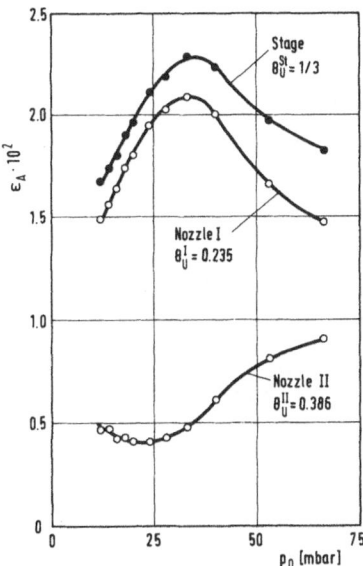

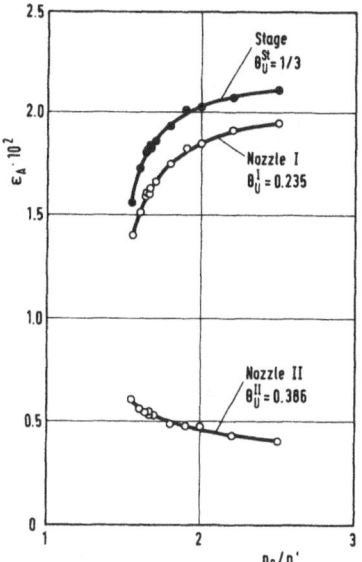

Fig.8.6. Influence of the inlet pressure p_0 upon the elementary effects of
isotope separation in the first nozzle
(ε_A^I), the second nozzle (ε_A^{II}), and the
stage (ε_A^{st}) at constant UF$_6$ cuts. Results of separation experiments using
an H$_2$/UF$_6$ mixture in a double-deflection system; $N_U^0 = 0.03$, $p_0/p' = 2.5$,
$p' = p''$, $T_0 = 296$ K; for the dimensions of the double-deflection system, see Fig.8.4

Fig.8.7. Influence of the expansion ratio p_0/p' upon the elementary effects
of isotope separation in the first noz-
zle (ε_A^I), the second nozzle (ε_A^{II}), and
the stage (ε_A^{st}) at constant UF$_6$ cuts.
Results of separation experiments using
an H$_2$/UF$_6$ mixture in a double-deflection
system; $N_U^0 = 0.03$, $p_0 = 24$ mbar, $p' = p''$,
$T_0 = 296$ K; for the dimensions of the
double-deflection system, see Fig.8.4

isotope separation process as given by the separation of UF$_6$ and auxiliary gas and
by the properties of the flow in the first nozzle. For simplification it may be assumed that the gas mixture entering the second nozzle is greatly slowed down in the
convergent channel constituted by the deflection wall and the skimmer of the first
nozzle and that, consequently, no major concentration gradients exist in the inlet
flow into the second nozzle. Isotope separation will then be governed mainly by the
UF$_6$ molar fraction and the average stagnation pressure of the inlet flow, because
these quantities determine the UF$_6$ speed ratio in the second nozzle at given suction pressures.

In the following sections, it is assumed that the average stagnation pressure of
the inlet flow in the second nozzle approximately agrees with the pressure which is
established in the second nozzle, if the valves of the intermediate and the heavy

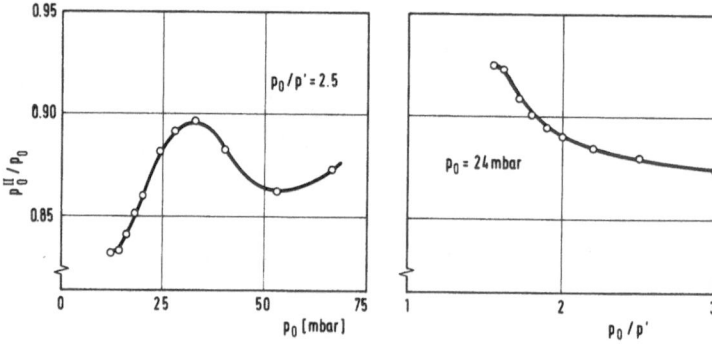

Fig.8.8. Influence of the inlet pressure p_0 and the expansion ratio p_0/p' upon the effective inlet pressure p_0^{II} of the second nozzle of a double-deflection system; for the experimental conditions, see Figs.8.6 and 8.7

fractions are closed, i.e., at $\theta_u^I = 1.$[22] This "effective" inlet pressure p_0^{II} of the second nozzle is plotted in Fig.8.8 as a function of the inlet pressure p_0 and the expansion ratio p_0/p' of the double-deflection system. The skimmer position of the first nozzle was always set to the value which, under normal operating conditions of the double-deflection system, results in the set value assumed here of $\theta_u^I = 0.235$ for the first nozzle. It is seen that p_0^{II} is some 10% lower on the average than the inlet pressure p_0 of the first nozzle. In the range of low inlet pressures, the ratio p_0^{II}/p_0 first increases with p_0, which can be explained by the decreasing viscous losses and the increasing concentration of UF_6 in the flow near the deflection wall. In the range of intermediate inlet pressures, the p_0^{II}/p_0 ratio decreases with increasing p_0, since the separation of UF_6 and auxiliary gas decreases and, hence, the concentration of UF_6 in the flow near the deflection wall becomes smaller. In the range of high inlet pressures, the p_0^{II}/p_0 ratio rises again because of the decreasing influence of viscous effects upon the separation nozzle flow. The decrease of the p_0^{II}/p_0 ratio with the expansion ratio is mainly due to the fact that the skimmer distance pertaining to $\theta_u^I = 0.235$ decreases with p_0/p' (Fig.6.4). Because of the radial decrease of velocity at the deflection wall, a reduction of the skimmer

---

[22] A more precise determination of the average dynamic pressure is possible by means of the molecular probe method which, unlike the pressure measurement in complete stagnation of the heavy fraction, does not change the flow field upstream from the skimmer. Since such measurements carried out for different operating pressures and skimmer positions are very expensive, and extrapolation of the results obtained on model gas mixtures to H_2/UF_6 mixtures may cause problems, only a few tentative studies of this type have so far been performed /68/. They revealed that the average dynamic pressure is slightly higher than the pressure measured at $\theta_u^I = 1$, which is also backed by the results of stagnation measurements shown in Fig.6.5.

distance is associated with a decrease of the dynamic pressure of the gas streaming into the channel formed by the skimmer and the deflection wall (Figs.5.6 and 6.5).

Since the differences described here between the inlet pressure p_0 of the first nozzle and the effective inlet pressure p_0^{II} of the second nozzle will always remain relatively small, the effective expansion ratios of the two nozzles are approximately equal. The Reynolds numbers of the flows in the two nozzles, which are determined mainly by the $\dot{UF_6}$ throughput, differ only slightly; at the given UF_6 cut of $\theta_0^I = 0.235$, the Reynolds number of the flow in the second nozzle is about 30% smaller than in the first nozzle. This allows the direct conclusion to be drawn that the average speed ratio of the mixture in the second nozzle is not considerably smaller than in the first nozzle, which has also been confirmed by measurements with free molecular probes /68/.

Unlike the speed ratio of the mixture, the partial speed ratio of the UF_6 in the second nozzle is considerably lower than in the first nozzle, because of the much higher UF_6 molar fraction N_u^{II} of the gas mixture fed to it. Since the separation factor of the UF_6/auxiliary gas mixture and, hence, the UF_6 molar fraction N_u^{II} vary greatly with the inlet pressure and the expansion ratio of the first nozzle, the boundary conditions for isotope separation in the second nozzle change strongly with the operating conditions of the first nozzle. This is exemplified by Fig.8.9, where the separation effect ε_A^{II} of the second nozzle and the UF_6 molar fraction N_u^{II} of the gas mixture fed to it are plotted versus the inlet pressure.

It is seen that ε_A^{II} and N_u^{II} change in opposite directions and that the minimum of ε_A^{II} agrees with the maximum of N_u^{II} and, accordingly, with the maximum separation be-

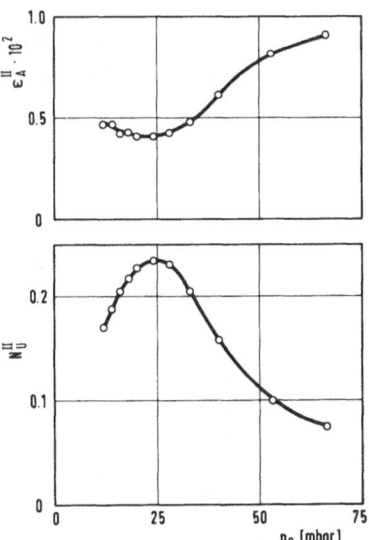

Fig.8.9. Influence of the inlet pressure p_0 upon the separation effect ε_A^{II} of the second nozzle of a double-deflection system and upon the UF_6 molar fraction N_u^{II} of the H_2/UF_6 mixture fed to the second nozzle; for the experimental conditions, see Fig.8.6

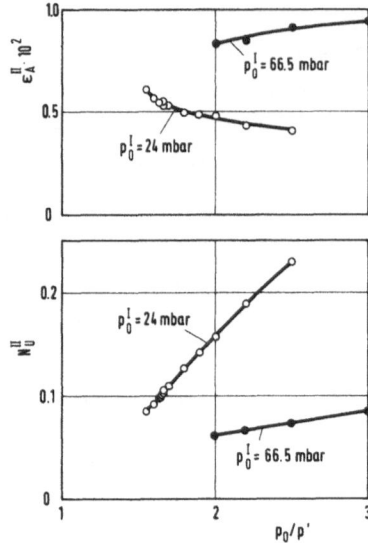

Fig.8.10. Influence of the expansion ratio p_0/p' upon the separation effect ε_A^{II} of the second nozzle of a double-deflection system and upon the UF_6 molar fraction N_u^{II} of the H_2/UF_6 mixture flowing into the second nozzle for various inlet pressures p_0; for the experimental conditions, see Fig.8.7

tween UF_6 and auxiliary gas in the first nozzle. A corresponding correlation between ε_A^{II} and N_u^{II} was also found in a variation of the expansion ratio evident from Fig. 8.10. At an inlet pressure of 24 mbar, which corresponds roughly to the value optimal for separation of the mixture ε_A^{II} decreases despite the increase in the effective expansion ratio p_0^{II}/p'. This is due to the fact that the increasing separation between UF_6 and auxiliary gas in the first nozzle causes the UF_6 molar fraction N_u^{II} of the feed stream to the second nozzle to rise steeply. The partial speed ratio of the UF_6 thus decreases in the second nozzle, although the speed ratio of the mixture increases. As is evident from the measurements at p_0 = 66.5 mbar, the increase in the speed ratio of the mixture dominates at high inlet pressures. In this case, N_u^{II} is clearly lower than at p_0 = 24 mbar, rising only relatively weakly with p_0/p'. Consequently, also ε_A^{II} is much higher than at 24 mbar, and there is a weak increase in ε_A^{II} with the expansion ratio.

8.2.3 Separative Work Output, Specific Process Parameters and Number of Separation Stages

Figure 8.11 is a plot of the inlet pressure optimal for isotope separation, p_0^+, the elementary effect ε_A^{St} of isotope separation for $p_0 = p_0^+$, and the net UF_6 throughput of the double-deflection system shown in Fig.8.4 versus the expansion ratio p_0/p'. As in the figures described above, the UF_6 cut of the first nozzle was set at θ_u^I = 0.235 and that of the second nozzle at θ_u^{II} = 0.386, resulting in a UF_6 stage cut of θ_u^{St} = 1/3. For comparison, the corresponding data are also indicated for the first nozzle of the double-deflection system, which is assumed to be operated as a single-

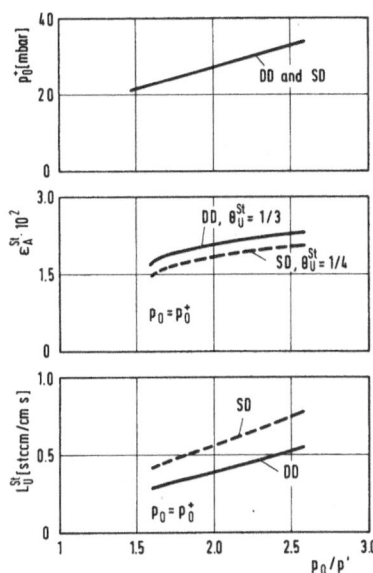

Fig.8.11. Optimum inlet pressure p_0^+, stage separation effect $\varepsilon_A^{St}(p_0=p_0^+)$, and net UF_6 throughput $L_u^{St}(p_0=p_0^+)$, per unit length of nozzle slit versus expansion ratio p_0/p' for a single-deflection system (SD) and a double-deflection system (DD). Results of separation experiments using an H_2/UF_6 mixture; $N_u^0 = 0.03$, $p'' = p'$, $T_0 = 293$ K; for the dimensions of the separation nozzles, see Fig.8.4

deflection system with $\theta_u = 1/4$. It is seen that the double-deflection and single-deflection systems do not differ greatly with respect to the optimum inlet pressure, which is self-evident because of dominating influence upon ε_A^{St} of the first nozzle (Fig.8.6). The absolute difference in separation effects of the double and single-deflection systems changes very little in the measurements shown here, i.e., the separation effect ε_A^{II} of the second nozzle is nearly independent of the expansion ratio at the inlet pressure p_0^+ optimal for isotope separation. The increase in the effective expansion ratio p_0^{II}/p' in the second nozzle, which is positive with respect to isotope separation, obviously is just compensated by the increase, adverse for isotope separation, in the UF_6 molar fraction, N_u^{II} of the gas mixture flowing into the second nozzle. Since the separation effect ε_A^I of the first nozzle rises with the expansion ratio for $p_0 = p_0^+$, contrary to ε_A^{II}, the contribution by the second nozzle to the overall separative work output δU^{St} of the double-deflection system becomes smaller and smaller as the expansion ratio becomes higher. Consequently, the separative work output of the single-deflection system can become higher, at high expansion ratios, than that of the double-deflection system, whose net UF_6 throughput according to (8.1) is about 30% lower than that of the single-deflection system under the UF_6 cuts assumed here of $\theta_u^I = 0.235$ and $\theta_u^{II} = 0.386$.

The technical expenditure for uranium enrichment by the double-deflection system is characterized by the specific process parameters in an analogous way as in the single-deflection system. The ideal specific energy consumption is defined, in accordance with (3.5), by the relation

$$E_s^{id} = L_m^V \; RT \; \ln(p_0/p')/\delta u^{St} \quad , \tag{8.6}$$

where L_m^V characterizes the throughput of the UF_6/auxiliary gas mixture through the compressor of the separation stage. In analogy with (3.6,7) the following relations apply to the ideal specific suction volume and the ideal slit length of the double-deflection system:

$$V_s^{id} = L_m^V \; RT/(p' \delta u^{St}) \tag{8.7}$$

and

$$l_s^{id} = 1/\delta u^{St} \quad . \tag{8.8}$$

For a more accurate comparison of the respective technical expenditures involved in double and single-deflection systems, the number of separation stages required for an enrichment plant must also be taken into account. As is seen in (3.10), the number of stages Z is inversely proportional to the elementary effect, $\varepsilon_\beta = \theta_u \cdot \varepsilon_A$, of depletion for separation nozzle cascades.

The specific process parameters and the reciprocal value of ε_β, which characterizes the number of stages, of the double-deflection system and the single-deflection system are plotted versus the expansion ratio in Fig.8.12 for $p_0 = p_0^+$. At low expansion ratios, in which the minimum of specific energy consumption is passed, the specific process parameters of the double-deflection system are some 10% lower than those of the single-deflection system. At higher expansion ratios, the differences in the specific process parameters of the two systems become smaller and smaller because of the decreasing contribution by the second nozzle to the overall separative work output of the double-deflection system.

In case of the double-deflection system, the number of separation stages in a cascade is more than 30% lower than in case of the single-deflection system, as is evident from the reciprocal values of ε_β. This reduction in the number of stages, brought about by the increase in the stage cut and in the separation effect per stage ε_A^{St}, can result in clear economic advantages. On the one hand, these are the reduced expenditures for instrumentation, which rise roughly proportional to the number of stages. On the other hand, the specific fabrication expenditure for a separation stage, i.e., the investment costs per unit output of separative work, decreases with increasing stage size. Since in plants of the same separative work output the separation stage is larger in case of the double-deflection system, but the number of stages is correspondingly smaller, there is a reduced overall expenditure for fabrication compared with plants using the single-deflection system.

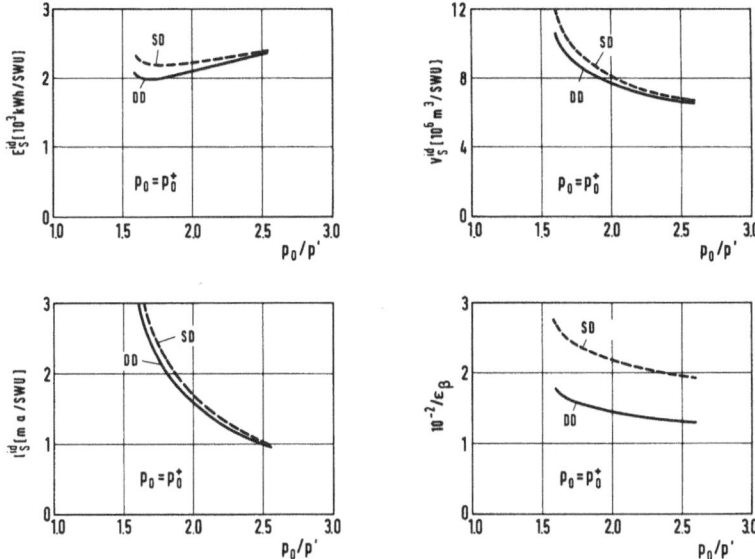

Fig.8.12. Specific energy consumption E_s^{id}, specific suction volume V_s^{id}, specific slit length l_s^{id}, and reciprocal depletion effect ε_β, at optimum inlet pressure p_0^+ for a single-deflection system (SD) and a double-deflection system (DD). Results of separation experiments using an H_2/UF_6 mixture; $\theta_u^I = 0.235$, $\theta_u^{II} = 0.386$, $\theta_u^{St} = 1/3$ (DD) and $\theta_u^{St} = 1/4$ (SD), $N_u^0 = 0.03$, $p'' = p'$, $T_0 = 293$ K; for the dimensions of the separation nozzles, see Fig.8.4

The advantages of the lower specific process parameters and the considerably smaller number of stages can be offset by the disadvantage of the higher expenditure involved in mass fabricating technical-scale double-deflection systems. Some tentative economic assessments indicate that on the basis of pessimistic assumptions with respect to the fabrication costs of technical double-deflection systems, the separative work costs would correspond to those of single-deflection systems. Under optimistic assumptions, savings of about 20% may be possible by using double-deflection systems in an enrichment plant.

8.3 Systems with Dynamic Flow Deflection

8.3.1 Basic Principles and Designs of Separation Systems

In the separation nozzle systems considered so far the separating centrifugal field is produced by deflecting a flow at a fixed, curved wall. Because of the velocity gradients in the flow region near the deflection wall, relatively high viscous losses occur in the curved flow. One fundamental possibility to reduce the velocity gradients in the flow region at the periphery of the centrifugal field is the prin-

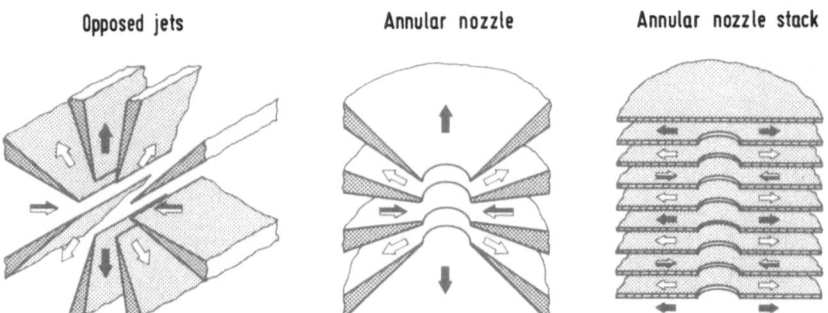

Opposed jets   Annular nozzle   Annular nozzle stack

Fig.8.13. Designs of separation nozzle systems with dynamic flow deflection

ciple of the so-called dynamic flow deflection /29,77-80/. In this method, the centrifugal forces are not acting upon a fixed wall, but the balance of forces in curved
flows, mirror symmetric to each other, is utilized.

The most important designs of separation nozzle systems with dynamic flow deflection are shown in Fig.8.13. In the opposed jet separation nozzle, the curvature of
the streamlines is generated by the mutual deflection of two band-shaped gas jets
emanating from two opposed, convergent nozzles. The mixture streaming away from both
sides of the plane connecting the nozzles is split by flow dividers arranged normal
to the nozzles into four light and two heavy partial streams.

In the rotationally symmetrical annular separation nozzle, the mixture to be separated is fed radially from the outside as a sink flow between two diaphragms shaped
like truncated cones. The flow is deflected in the region surrounding the axis of the
system and removed in opposite directions through the openings of the two conical diaphragms. Two other conical diaphragms are used to split the flows into light and heavy
fractions which are withdrawn radially to the outside and along the axis of rotation,
respectively. As a result of the radial increase in flow cross sections, the flow velocities decrease quickly as the distance from the axis increases, so that only low
viscous losses occur in the incoming and outgoing gas flows.

This radial expansion of the flow cross sections even makes it possible to replace
the conical diaphragms by plane diaphragms without causing drastic increases in viscous losses of the gas flows passing between the diaphragms. In this way, a large
number of individual annular separation nozzles can be combined in a compact nozzle
stack in which alternately the mixture to be separated is fed and the light and heavy
partial streams are extracted between the diaphragms. Since all nozzles are arranged
on a common axis, the heavy partial streams of two adjacent nozzles coming from two
opposite directions may be passed into a common collection chamber and deflected additionally by the opposed jet principle.

In the technical implementation of the separation nozzle method, the systems employing dynamic flow deflection have not yet achieved any significance so far. There are two main reasons for this. On the one hand, the fabrication techniques existing at present for technical-scale separation nozzle elements cannot easily be utilized for producing those arrangements. On the other hand, the systems with dynamic flow deflection so far have shown specific process parameters of comparably favorable levels only at low expansion ratios, as will be explained in more detail below. Accordingly, the elementary effect of isotope separation is small under optimal operating conditions of the process, which means that the number of separation stages in an enrichment plant must be correspondingly large. The use of low-cost large separation stages is bound to result in plants with a correspondingly high separation capacity, which would unneccessarily augment the financial risk associated with the industrial implementation of the separation nozzle process. However, for the long-term development of the separation nozzle process, the systems employing dynamic flow deflection may become quite attractive. The development work performed in this field, which will be reported below, therefore mainly dealt with the specific physics phenomena of dynamic flow deflection, while the optimization of geometric and operating parameters has been limited so far to a few experiments only.

8.3.2 Interaction of Opposed Jets at Medium Knudsen Numbers

For studies of the flow field and of the spatial development of separation in systems employing dynamic flow deflection, the free molecular probe method and, to a lesser extent, flow diagnostics by means of a CO_2 laser have been applied (Sect. 5.3.1). Below, reference will be made mainly to measurements on the opposed jet separation nozzle, because it has been studied in much more detail than the annular separation nozzle. The coordinates of the opposed jet separation nozzle used to represent the measurements and the nomenclature of the geometric parameters can be seen in Fig.8.14.

Figure 8.15 shows the local flow directions of the heavy component of the mixture as determined by molecular probe measurements on an opposed jet system operated with a He/SF$_6$ mixture; the gas stream of the heavy fraction in this case was throttled very strongly. It is seen that there is an extended stagnation zone around the center of the separation element. The gas flows emanating from the two nozzles are deflected by the stagnation and, corresponding to the symmetry of the system, four individual stationary centrifugal fields are formed, whose forces balance each other. The stream surfaces characterized by the planes of symmetry of $x = 0$ and $y = 0$, which correspond to the peripheries of the four centrifugal fields, can have an angle of deflection of 90° only. However, the inner stream surfaces have an angle

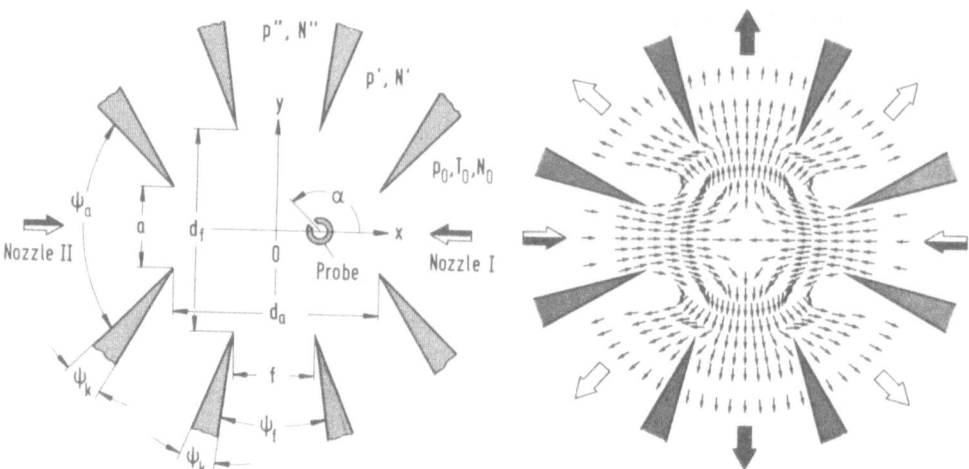

Fig.8.14. Coordinates and geometric parameters of the opposed jet separation nozzle

Fig.8.15. Local flow directions of the heavy component of the mixture. Results of free molecular probe measurements using a He/SF$_6$ mixture in an opposed jet separation nozzle; p_0 = 0.53 mbar, N_s^0 = 0.04, p_0/p' = 2, p_0/p'' = 1.3, a = 20 mm, f = 30 mm, d_a = 60 mm, d_f = 60 mm, ψ_a = 30°, ψ_f = 30°, ψ_k = 15°

of deflection of almost 180°, which is advantageous for isotope separation because of the longer radial diffusion path of the inner stream surfaces (Sect.8.1.1). Stagnation of the heavy fraction causes the curvature of the stream surfaces to be increased in the region right before the skimmer and the angle of deflection to become larger.

The Knudsen numbers optimal for isotope separation, which are usually referred to the nozzle width a in an opposed jet separation nozzle, are in the range of 0.05 to 0.01, as in a standard separation nozzle. Consequently, there are major changes of state within few mean free paths of the molecules, especially in the stagnation region of the flow, which results in pronounced non-equilibrium effects (Sect.5.3.3). More precise analysis of the free molecular probe measurements in particular shows that the opposed gas jets partly penetrate into each other /33,81/.

This phenomenon is illustrated by Fig.8.16, which is a plot of the ion current as a function of the angle of attack of the probe opening for the heavy component of the mixture; the probe was located in the center of the separation element (x = 0, y = 0), and in the plane connecting the nozzles (x/a = 0.133, y = 0); the test gas used was a He/C$_7$F$_{14}$ mixture. In the center of the separation element, the ion current curve has two peaks of the same height, which are shifted relative to each other by 180°, the probe opening in each case facing in the direction of one of the two nozzles. With increasing distance from the center, the maximum to be assigned

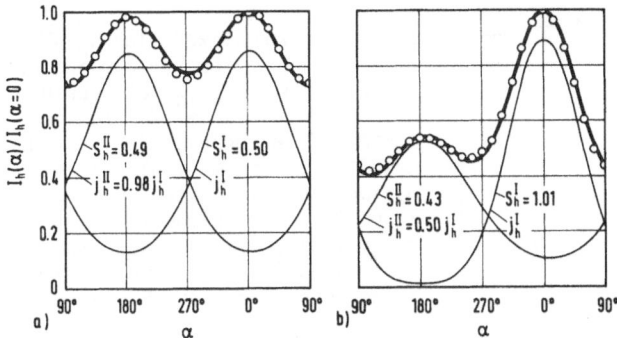

Fig.8.16. Normalized C_7F_{14} ion current $I_h/I_h(\alpha=0)$ versus angle of attack α of the opening of a free molecular probe in the stagnation zone of an opposed jet system operated with a He/C_7F_{14} mixture. The curves drawn through the measured points were calculated for a bimodal superposition of Maxwellian distributions attributed to the opposed jets penetrating into each other with the speed ratios S_h^I, S_h^{II} and the fluxes j_h^I, j_h^{II}. The thin curves indicate the hypothetical variations of the ion current for the individual Maxwellian distributions. Probe coordinates (see Fig.8.14): (a) $x/a = 0$, $y/a = 0$; (b) $x/a = 0.133$, $y/a = 0$. Operating conditions: $p_0 = 0.067$ mbar, $p_0/p' = 2.33$, $N_h^0 = 0.04$. Geometric parameters (see Fig.8.14): $a = 30$ mm, $d_a = 75$ mm, $\psi_a = 90°$, $\psi_k = 10°$; no skimmers were installed for these measurements

to the more distant nozzle becomes smaller and smaller and finally disappears at the periphery of the stagnation zone. The curves drawn through the measured points were calculated under the assumption that the local velocity distribution of the molecules in accordance with (5.16) can be regarded as a superposition of two Maxwellian distributions, which must be assigned to the opposed gas jets partly penetrating into each other. Within the framework of this model of a bimodal molecular velocity distribution, which has been discussed in detail in Sect.5.3.3, separate values of the partial speed ratios and fluxes can be determined within the penetration zone for the two jets. Accordingly, it follows from the measurements shown in Fig.8.16 that the C_7F_{14} speed ratios are about 0.5 for the opposed jets in the center of the separation element.

The separated fluxes and speed ratios of the two jets are plotted in Figs.8.17,18 versus the distance x/a from the center of the separation element. In order to indicate clearly the effect of mutual penetration, these measurements were performed at a relatively high Knudsen number of $Kn \cong 0.1$. It is seen that the C_7F_{14} flux reaches a maximum about half a nozzle width downstream from the nozzle opening; at the center of the separation element, the C_7F_{14} flux attributed to each single jet amounts to about 45% of the maximum value. In an analogous way, the speed ratio of the gas jet emanating from the nozzle first increases, reaching a maximum value of $S_h = 2.7$ for the heavy component of the mixture, while the subsequent deceleration of the flow results in a decrease of S_h to a value of 0.5 at the center of the separation element. The flux of the molecules penetrating the $x = 0$ plane drops quickly, as is

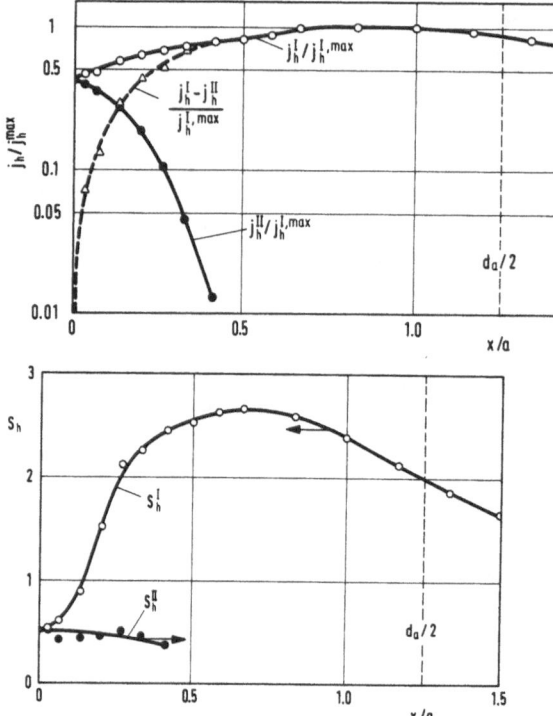

Fig.8.17. Normalized flux j_h/j_h^{max} of C_7F_{14} along the x = 0 plane of two opposed He/C_7F_{14} jets versus normalized distance x/a from the center of the opposed jet system. Results of free molecular probe measurements evaluated under the assumption of a bimodal molecular velocity distribution within the stagnation zone; for the operating conditions, see Fig.8.16

Fig.8.18. Speed ratio S_h of C_7F_{14} along the y = 0 plane of two opposed He/C_7F_{14} jets versus normalized distance x/a from the center of the opposed jet system. Results of free molecular probe measurements evaluated under the assumption of a bimodal molecular velocity distribution within the stagnation zone; for the operating conditions, see Fig.8.16

evident from the mirror symmetrical curve of the flux of the opposed jet plotted in Fig.8.17. At a distance of half a nozzle width from the center of the separation element, the flux of the molecules penetrated through the x = 0 plane is about two orders of magnitude lower than at the center of the separation element. The speed ratio of the molecules penetrating into the opposed jet decreases only weakly, which is in contrast to the strong decrease of the flux. This can be explained by the fact that molecules with a high-velocity component in the direction of the opposed nozzle penetrate more deeply into the opposed jet and those with a high transverse velocity component experience a larger number of collisions. Conditions qualitatively similar to the measurements described here for the opposed jet system can be found in the annular separation nozzle, in which a penetration zone is formed in the center of the sink flow of the nozzle feed duct.

At the Knudsen numbers optimal for separation, the flux of the molecules penetrating through the x = 0 plane is still some 10% to 20% of the maximum flux. Hence, it must be assumed that isotope separation, both in the opposed jet and in the annular separation nozzles, is influenced greatly by the interpenetration of the opposed flows and that mixing of flow regions of different isotopic compositions occurs in the stagnation zone. In addition, kinetic energy of directional opposed motion is irreversibly dissipated in the penetration zone, which gives rise to pressure losses and, accordingly, to a reduction of the separating centrifugal forces. These effects, which are detrimental to isotope separation, become more and more pronounced with increasing expansion ratio; in a qualitative way, this explains the unfavorable separation characteristics of the systems with dynamic flow deflection at high expansion ratios mentioned in Sect.8.3.1.

8.3.3 Acceleration of Disparate Mass Mixtures in Convergent Nozzles

In the systems with dynamic flow deflection, as in those with mechanical jet deflection, isotope separation is affected positively by the UF_6 molar stream surfaces being as far as possible away from their equilibrium positions at the beginning of deflection (Sect.8.1.1). This prerequisite for a high separation effect can be met in the opposed jet system by making the inlet angle ψ_a of the convergent slit-shaped nozzles as large as possible. In this case, a trapezoidal velocity profile is generated in the range of the nozzle opening, which is typical of a strongly accelerated flow, while a small inlet angle results rather in the formation of a parabolic velocity profile typical of a fully developed viscous flow. When regarding the four separated centrifugal fields of the opposed jet flow (Fig.8.15), it is evident that the parabolic flow profile has the UF_6 molar stream surfaces relatively close to their equilibrium positions; the UF_6 is concentrated near the peripheries of the centrifugal fields at the beginning of deflection, since the UF_6 flux has a pronounced maximum at the middle of the nozzle opening. Compared with a parabolic flow profile, a trapezoidal profile obviously results in a longer radial diffusion path of the UF_6 stream surfaces in the auxiliary gas. The most favorable inlet angle for isotope separation, as was found in separation experiments, is around 75°; increasing ψ_a further then impairs the separation characteristics again, since the suction conditions for the light and heavy fractions become more and more adverse.

The rapid acceleration of the mixture at large inlet angles results in major changes of state within a few mean free paths in the flow region near the nozzle opening. In this region, the light auxiliary gas is accelerated more strongly than the heavy component of the mixture, and a local velocity slip is produced between the light and heavy molecules, to which reference has already been made in Sect.5.3.3. This phenomenon is evident from the results shown in Fig.8.19 of free molecular probe measurements, in which the partial speed ratios and the partial pressures were determined for a He/C_7F_{14} mixture in a rapidly converging nozzle /66/. It is seen that the speed ratio S_h of the heavy component clearly rises more slowly in the acceleration of the mixture than does the speed ratio S_a of the auxiliary gas multiplied by the square root of the molecular weight ratio $\sqrt{M_h/M_a}$. Since the throughput of the heavy component of the mixture has the same value for each cross section of the

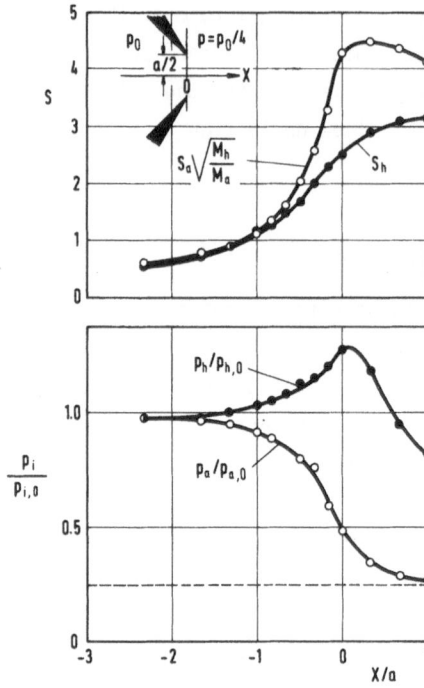

Fig.8.19. Partial speed ratios S_h and S_a of the heavy component and the light auxiliary gas, and partial pressures p_h and p_a normalized to their reservoir values versus normalized distance X/a from the opening of a convergent slit-shaped nozzle. Results of free molecular probe measurements using a He/C_7F_{14} mixture in a single nozzle; $N_h^0 = 0.04$, $p_0 = 0.13$ mbar, $p_0/p = 4$, $a = 30$ mm, $\psi_a = 90°$

steady-state flow, its lower velocity must be off-set by a change in partial density.[23] The velocity slip even causes an increase in partial pressure of the heavy component, while the partial pressure of the light auxiliary gas decreases corresponding to the expansion of the mixture.

This immediately gives rise to the conclusion that the increase in the inlet angle which has a positive impact on the UF$_6$ distribution at the beginning of deflection, also entails a mechanism negative for isotope separation. For, if the mixture is accelerated rapidly within a few mean free paths, the acceleration of the UF$_6$ by the auxiliary gas is less effective, and the centrifugal forces are reduced. Moreover, a velocity slip is always connected with additional entropy generation and, hence, with increased pressure losses /67/.

---

[23] As was outlined in Sect.5.3.1, free molecular probe measurements can be applied only to determine speed ratios, not absolute values of flow velocities of the components of the mixture. In order to supplement molecular probe measurements, additional studies were performed by the laser method (Sect.5.3.1) which allows both the partial density and the rotational/vibrational temperature of the heavy component of the mixture to be determined /62,63/. It follows from these additional measurements that the differences observed between S_h and $S_a \cdot \sqrt{M_h/M_a}$ are mainly due to a lower flow velocity of the heavy component of the mixture.

As has been mentioned in Sect.5.3.3, the velocity slip encountered under conditions of rapid acceleration of the mixture considerably influences the throughput through the nozzle /66/. The throughput behavior is characterized below by the so-called discharge coefficient,

$$C_D = \dot{m}_{exp}/\dot{m}_{is}^* \quad , \tag{8.9}$$

with $\dot{m}_{exp}$ being the mass flow measured and $\dot{m}_{is}^*$ the maximum mass flow calculated for a quasi one-dimensional isentropic expansion. It is well known that

$$\dot{m}_{is}^* = F^* \rho^* c^* = F^* \rho_0 c_0 \left[2/(\gamma+1) \right]^{\frac{\gamma+1}{2(\gamma-1)}} \quad , \tag{8.10}$$

with F^* as the section of minimum area (throat) of the nozzle, ρ^* the density and c^* the sonic velocity at the throat; ρ_0 is the density and c_0 the sonic velocity under reservoir conditions, γ is the ratio of the specific heats. Under isentropic flow conditions, the discharge coefficient is

$$C_{D,is} = Ma \left[1 + \frac{\gamma-1}{\gamma+1} (Ma^2-1) \right]^{-\frac{\gamma+1}{2(\gamma-1)}} \quad , \text{ for } Ma \leq 1 \tag{8.11a}$$

and

$$C_{D,is} = 1 \quad , \text{ for } Ma \geq 1 \quad , \tag{8.11b}$$

where the Mach number, Ma, of the expanding gas follows from the well known relation

$$Ma^2 = \frac{2}{\gamma-1} \left[(\frac{p_0}{p})^{\frac{\gamma-1}{\gamma}} - 1 \right] \quad . \tag{8.12}$$

In Fig.8.20, the discharge coefficient C_D has been plotted for an H_2/UF_6 mixture and for a uniform gas (N_2) versus the expansion ratio p_0/p applied to the convergent nozzle. At high expansion ratios, the Reynolds number of the flow is about 100. The dependence on the expansion ratio of the discharge coefficient $C_{D,is}$ of an isentropic flow is described by the dashed curve. It is seen that the discharge coefficient of the H_2/UF_6 mixture considerably exceeds that of the uniform gas and even assumes values in excess of 1 at high expansion ratios; the discharge coefficient of the uniform gas remains clearly below the isentropic value $C_{D,is}$ as was to be expected because of the low Reynolds number of the flow.

This surprising behavior of the mixture can be explained by the fact that the incomplete momentum transfer between the light and heavy molecules increases the speed of propagation of disturbances in the mixture according to the higher thermal veloci-

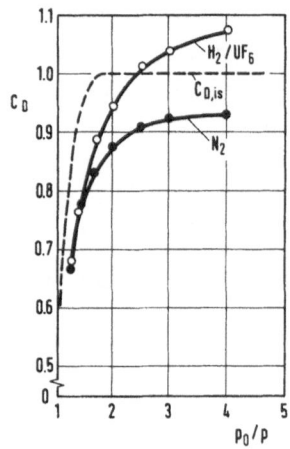

Fig.8.20. Discharge coefficient C_D of a convergent slit-type nozzle versus expansion ratio p_0/p for an H_2/UF_6 mixture and for a uniform gas (N_2) at Reynolds numbers around 100. The dashed curve indicates the dependence on the expansion ratio of the discharge coefficient $C_{D,is}$ for an isentropic flow with a ratio of the specific heats of $\gamma = 1.4$. Results of mass flow and gas meter measurements in a single nozzle; $p_0 = 32$ mbar for H_2/UF_6 and N_2, $N_u^0 = 0.04$, $a = 0.3$ mm, $\psi_a = 90°$

ty of the lighter molecules. Thus, the effective sonic velocity becomes higher, if the pressure changes strongly within few mean free paths. Since this phenomenon occurs only under conditions of rapid acceleration in a rapidly converging nozzle, the opposed jet system fundamentally differs from the standard separation nozzle in terms of its throughput behavior, because the configuration of the latter permits only a comparatively slow expansion of the flow. In the standard separation nozzle, the discharge coefficients for disparate mass mixtures are clearly even lower than for uniform gases if identical Reynolds numbers of the flow and identical expansion ratios are assumed. This lower discharge coefficient of the mixture can be explained by pressure losses resulting from the separation between UF_6 and auxiliary gas (Fig.5.13).

8.3.4 Flow Stability of Opposed Jets

If the Reynolds number of the flow in an opposed jet nozzle is increased beyond a certain value, the symmetrical, steady-state flow configuration can suddenly change into a flow configuration unsuitable for separation /77/. In separation experiments, this flip-over is indicated by a sudden decrease of the elementary effect ε_A of isotope separation and the separation factor A_m of the mixture and in an increase in the UF_6 cut θ_u if the inlet pressure p_0 exceeds a critical level. Accordingly, under conditions of decreasing inlet pressure, one observes a sudden increase in ε_A and A_m accompanied by a decrease of θ_u. As is seen from the measurements shown in Fig.8.21, the level of the inlet pressure at which the sudden change in flow configuration occurs depends on the sense in which the pressure changes. The critical inlet pressure is higher if one starts at low inlet pressure, i.e., at the symmetrical, steady-state flow configuration appropriate to separation. The variation of ε_A, A_m and θ_u with the inlet pressure thus has a hysteresis-type behavior.

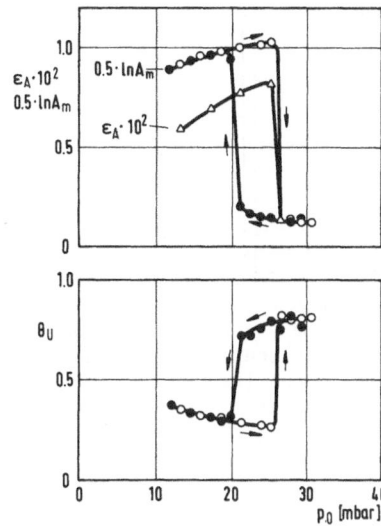

<u>Fig.8.21.</u> Variation of the elementary effect ε_A of isotope separation, of the separation factor A_m of the mixture, and of the UF_6 cut θ_u, with the inlet pressure p_0 in the flip-over range of opposed jets. Results of separation experiments using an H_2/UF_6 mixture in an opposed jet separation nozzle; $N_u^0 = 0.05$, $p_0/p' = 4$, $p'' = p'$, $a = 0.2$ mm, $d_a = 0.8$ mm, $f = 0.2$ mm, $d_f = 0.8$ mm, $\psi_a = 30°$, $\psi_f = 30°$, $\psi_k = 15°$

The critical Reynolds number for the flip-over of the flow configuration not only depends on the geometry of the separation system, but also on the expansion ratio p_0/p' of the light fraction, the stagnation ratio p''/p' and on the composition of the mixture. With decreasing expansion ratio p_0/p', the critical Reynolds number in general clearly increases, and at low expansion ratios no flip-over of the flow is observed at all. The critical Reynolds number of H_2/UF_6 mixtures is significantly lower than that of He/UF_6 mixtures; in model experiments using uniform gases and mixtures, whose components differed less strongly in terms of molecular weight (He/SF_6, N_2/C_7F_{14}, Ar/C_7F_{14}) the flow configuration was stable even at Reynolds numbers above 1000.

Model experiments performed on a He/C_7F_{14} mixture by means of laser flow diagnostics (Sect.5.3.1) showed that in the flip-over range the discharge of gas from one of the two nozzles is partly blocked by the respective opposed jet. This flow configuration is preserved only for a certain period of time and then suddenly changes into partial blocking of the other nozzle. This behavior is evident from Fig.8.22, where the partial density ν_h of the heavy component of the mixture is shown for two different Reynolds numbers along the plane connecting the two nozzles. Up to a Reynolds number of $Re \leq 250$, the curve is symmetrical to $x/a = 0$; it reflects the decrease of density associated with the expansion of the gas mixture in the nozzles and the high C_7F_{14} density in the center of the opposed jet system $\nu_h(x/a=0)$, which clearly exceeds the reservoir value ν_h^0. At higher Reynolds numbers ($Re \geq 250$), the partial density curve becomes asymmetrical. The very weak maximum of ν_h is shifted right in front of one of the two nozzle openings, so that the gas jet discharged by that nozzle opening expands only very weakly, while the oppo-

118

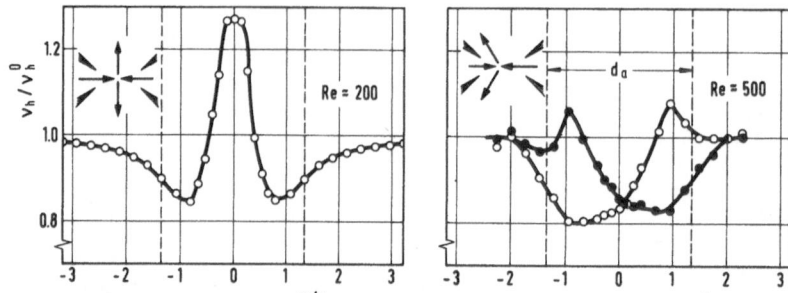

Fig.8.22. Normalized partial density ν_h/ν_h^0 of C_7F_{14} along the y = 0 plane of two opposed He/C_7F_{14} jets versus normalized distance x/a from the center of the opposed jet system for different Reynolds numbers Re of the flow. The partial density profiles at Re $\cong$ 500 appear alternately. The arrows schematically show the mean directions of the inflow and of the deflected flow. Results of laser flow diagnostics; N_h^0 = 0.04, p_0 = 2.67 mbar (Re $\cong$ 200) and p_0 = 6.67 mbar (Re $\cong$ 500), p_0/p' = 1.5, a = 7.5 mm, d_a = 18.3 mm, ψ_a = 75°, ψ_k = 10°; no skimmers were installed for these measurements

site jet has a much more extended and much more pronounced decrease of density. This density profile changes suddenly after a certain period of time, and an equivalent profile is developed, in which the maximum of the stagnation pressure is moved right in front of the other nozzle opening. The mean direction of outflow from the stagnation zone is no longer normal to the plane connecting the two nozzles, but the gas preferably flows out in the original direction of flow of the more strongly expanded gas jet, as is indicated schematically in Fig.8.22 by the arrows characterizing the mean flow directions.[24]

This flip-over of the flow configuration, in which the opposed jet system behaves like an unstable fluidic multivibrator, is probably influenced by the Coanda effect characterizing the attachment of a flow at a solid wall. In what way this attachment of the flow to the outer nozzle wall is supported by the separation of the mixture components and to what extent secondary flows in the boundary layers at the end walls of the slit-type system are responsible for the alternating flip-over, cannot be seen from these model experiments.

The instability of flow at high Reynolds numbers, however, does not jeopardize the practical application of opposed jet separation nozzles for uranium isotope separation. Nozzle arrangements which resulted in the most favorable values of the specific process parameters mostly had the flip-over of the flow only at operating condi-

---

[24] Besides the flow configuration described here, a configuration is also likely to occur at high Reynolds numbers in which the opposed gas jets slide on each other under a flat angle /77/.

tions deviating relatively strongly from the values attributed to the optimum operating point. In general, the inlet pressures, expansion ratios and UF_6 cuts characteristic of flow flip-over were approximately a factor of 2 above the values at which the minimum of specific energy consumption was obtained.

In contrast to the opposed jet separation nozzle, no flip-over in the flow configuration was observed in the annular separation nozzle. As is seen from Fig.8.23, the elementary effect ε_A of isotope separation in an annular separation nozzle, in analogy to the standard separation nozzle, decreases approximately with $1/Re$, while there is a sudden decrease of ε_A in the opposed jet separation nozzle.

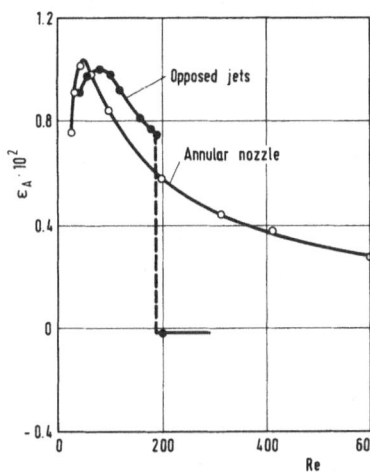

Fig.8.23. Typical dependences of the elementary effect ε_A of isotope separation on the Reynolds number Re of the flow for opposed jet and annular separation nozzles. Results of separation experiments using an H_2/UF_6 mixture; $N_u^0 = 0.04$, $p_0/p' = 1.5$; the UF_6 cut was set to a constant value of $\theta_u = 0.25$ by throttling of the heavy fraction

8.3.5 Separation of Uranium Isotopes in an Opposed Jet Separation Nozzle

The separation experiments performed so far with opposed jet separation nozzles mainly concentrated on systems built up of simple, wedge-shaped parts (Fig.8.14). A few experiments were also performed on systems in which the nozzles were rounded in the region of the opening; under comparable experimental conditions, however, those systems resulted in clearly lower values of the elementary effect of isotope separation. In accordance with the explanations given in Sect.8.3.3, this can be ascribed to the formation of a parabolic velocity profile in the region of the nozzle opening and, hence, to adverse initial conditions for isotope separation.

Figure 8.24 is a plot of the results of a typical series of experiments performed on an opposed jet separation nozzle using an H_2/UF_6 mixture at various expansion ratios p_0/p' of the light fraction. The inlet pressure p_0 and the skimmer width f were kept constant; the UF_6 cut was set to $\theta_u = 1/4$ by throttling of the heavy frac-

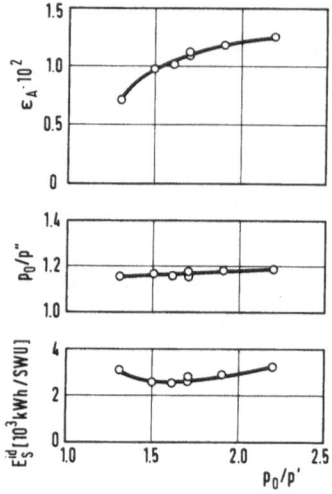

Fig.8.24. Elementary effect ε_A of isotope separation, expansion ratio p_0/p'' of the heavy fraction for setting the UF_6 cut to $\theta_u = 1/4$, and specific energy consumption E_s^{id} versus expansion ratio p_0/p' of the light fraction. Results of separation experiments using an H_2/UF_6 mixture in an opposed jet separation nozzle; $N_u^0 = 0.04$, $p_0 = 27$ mbar, $a = 0.3$ mm, $f = 0.28$ mm, $d_a = 0.75$ mm, $d_f = 0.8$ mm, $\psi_a = 90°$, $\psi_f = 14°$, $\psi_k = 10°$

tion, i.e., by increasing the suction pressure p" of the heavy fraction over the suction pressure p' of the light fraction.

Comparison with the corresponding measurements performed on a standard separation nozzle shows that the ε_A values of the opposed jet system are just as high as those of a standard separation nozzle up to expansion ratios of $p_0/p' \leq 1.6$; at higher expansion ratios, e.g. $p_0/p' > 2$, the ε_A values of the standard separation nozzle clearly exceed those of the opposed jet system. This may be explained by the fact that the opposed gas jets penetrate into each other more and more intensively with increasing expansion ratio. In addition, the acceleration of UF_6 by the auxiliary gas becomes less effective, because the velocity slip between the components of the mixture increases with the expansion ratio (Sects.8.3.2,3).

There is a clear difference between the opposed jet separation nozzle and the system with mechanical flow deflection in respect of the behavior under conditions of stagnation of the heavy fraction. The stagnation ratios p"/p' optimal for isotope separation are clearly higher in the opposed jet separation nozzle at a given expansion ratio p_0/p' of the light fraction. Furthermore, the relative increase in the separation effect, which can be achieved by stagnation of the heavy fraction at a constant UF_6 cut, is usually higher than in the standard separation nozzle (Sect.6.2). If the heavy fraction stream is throttled very strongly, the static pressure p" in the heavy fraction at high UF_6 cuts may even rise above the inlet pressure p_0, while in systems with mechanical flow deflection p" always remains smaller than p_0. These different kinds of behavior can mainly be explained by the fact that in an opposed jet separation nozzle the stagnation pressure of the heavy fraction stream rises continuously with increasing θ_u; the maximum stagnation pressure is attained at the periphery of the centrifugal field, i.e., at the plane of symmetry characterized by x = 0 (Fig.8.14). When the flow is deflected by a curved wall, the maximum stagnation pressure is not attained at the periphery of the centrifugal field ($\theta_u = 1$) because of the radial decrease of the flow velocity in the region close to the wall. The pressure losses in the heavy fraction thus are clearly lower in dynamic flow deflection than in flow deflection at a fixed wall.[25] Making use of the high stagnation pressure at the periphery of the centrifugal field (x = 0) is limited by the lower stagnation pressure at the edges of the skimmer. A backflow from the heavy fraction into the light fraction may occur, if the suction pressure of the heavy fraction p" becomes higher than the local stagnation pressure of the gas flowing towards the skimmer edges. This effect, which is bound to result in mixing losses, plays an important role mainly in the annular separation nozzle and will be discussed in more detail in the following section.

The optimum inlet pressure p_0^+ of the opposed jet separation nozzle rises, as in the systems with mechanical jet deflection, if the expansion ratio is increased or the UF_6 molar fraction is decreased (cf. e.g., Sects.6.1,4). At the same nozzle widths, the optimum inlet pressures of the opposed jet separation nozzle and the standard separation nozzle are nearly identical for low expansion ratios. The smaller angle of deflection of the UF_6 molar stream surfaces in the opposed jet separation nozzle is obviously compensated by the fact that the gas flow discharged from each nozzle forms two jets curved in opposite senses. The reference dimension for the Knudsen number of these jets is half the nozzle width and, accordingly, the optimum value Kn^+ is twice as high as in the standard separation nozzle. Nevertheless, the product of Kn and the angle of deflection ϕ of the outer stream surfaces of the centrifugal field, which determines the azimuthal position of the transient maximum of ε_A, remains unchanged (Sect.4.2.3).

At low expansion ratios of the light fraction ($p_0/p' \lesssim 1.6$ for $N_u^0 = 0.04$), the specific energy consumption of the opposed jet separation nozzle is significantly below that of the standard separation nozzle. This is due to the fact that at about the same values for ε_A the suction pressure of the heavy fraction in the opposed jet separation nozzle can be raised to a higher level by stagnation. If the opposed jet separation nozzle is operated, e.g., with an H_2/UF_6 mixture with a UF_6 molar

---

[25] This is not only the result of a comparison with the stagnation behavior of the standard separation nozzle, but also follows directly from separation experiments performed on an arrangement in which one of the two opposed jets was simulated by a plane wall arranged normal to the mean direction of discharge from the nozzle. In this arrangement, in which the curved flow is slowed down at the fixed wall, clearly lower values were found for the elementary effect of isotope separation and the stagnation pressure of the heavy fraction than in the corresponding opposed jet separation nozzle.

fraction of N_u^0 = 0.04, the minimum of specific energy consumption is passed at an expansion ratio of p_0/p' ≅ 1.6, while in a standard separation nozzle the minimum of E_s^{id}, using the same mixture, is reached at an expansion ratio of p_0/p' ≅ 2.1. The minimum values of E_s^{id} in both arrangements are about 2500 kWh/SWU. At low expansion ratios, the ideal specific slit length of the opposed jet separation nozzle is much smaller than that of the standard separation nozzle, because the gas is discharged from two nozzles which, in addition, have a much higher discharge coefficient. If, however, the two systems are compared at the expansion ratios pertaining to the minimum of E_s^{id}, the resultant values for the specific slit length are nearly identical. The specific suction volume of the standard separation nozzle in that case is only about 50% and the number of separation stages is about 70% of the corresponding values of the opposed jet system. Accordingly, at the present state of development, the overall technical expenditure for uranium isotope separation by the opposed jet separation nozzle is considerably higher than that of the standard separation nozzle.

8.3.6 Separation of Uranium Isotopes in an Annular Separation Nozzle

Figure 8.25 is a plot of the results of a typical series of measurements performed on an annular separation nozzle built up of conical nozzle diaphragms (Fig.8.13). The system was operated on an H_2/UF_6 mixture with a UF_6 molar fraction of 0.04, the inlet pressure and the geometric parameters of the system were kept constant, and the UF_6 cut was set at θ_u = 1/4 by throttling of the heavy fraction stream. The dia-

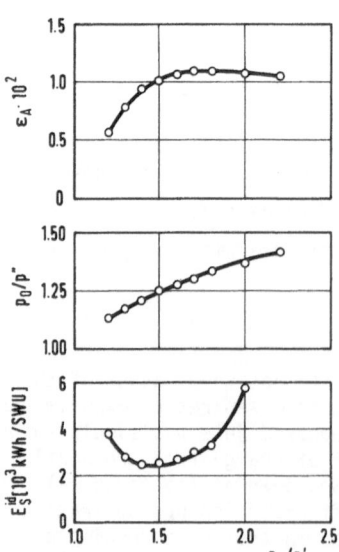

Fig.8.25. Elementary effect ε_A of isotope separation, expansion ratio p_0/p'' of the heavy fraction for setting the UF_6 cut to θ_u = 1/4, and specific energy consumption E_s^{id} versus expansion ratio p_0/p' of the light fraction. Results of separation experiments using an H_2/UF_6 mixture in an annular separation nozzle; N_u^0 = 0.04, p_0 = 4 mbar, $D_a = D_f$ = 2 mm, B_a = 0.8 mm, B_f = 0.95 mm

meters D_a of the openings in the nozzle diaphragms and the diameters D_f of the openings in the skimmer diaphragms were 2 mm, the width B_a of the annulus formed by the nozzle diaphragms was set at 0.8 mm, and the width B_f of the annulus formed by the nozzle and skimmer diaphragms was set at 0.95 mm.

It appears from the measurements that the elementary effect ε_A of isotope separation in this arrangement reaches a maximum at an expansion ratio of $p_0/p' \cong 1.7$. The ideal specific energy consumption passes through a minimum at $p_0/p' \cong 1.4$, the minimum value of E_s^{id} of 2500 kWh/SWU roughly corresponding to that of the opposed jet separation nozzle and the standard separation nozzle. The decrease of ε_A with the expansion ratio, which occurs for $p_0/p' > 1.7$, cannot be avoided by a further increase in the inlet pressure. Consequently, the occurrence of the maximum of ε_A at $p_0/p' \cong 1.7$ cannot be explained by the fact that for higher expansion ratios the maximum of isotope separation is shifted towards smaller angles of deflection according to the increase in UF_6 speed ratio (Sect.4.2.3).

In order to illustrate further the separating characteristics of the annular separation nozzle, the results of measurements with an H_2/UF_6 mixture with a relatively low UF_6 molar fraction of $N_u^0 = 0.02$ are shown in Fig.8.26. The UF_6 cut was set to $\theta_u = 1/2$ and $\theta_u = 1/3$ by stagnation of the heavy fraction. Unlike the measurements shown in Fig.8.25, the diameter of the openings in the skimmer diaphragms ($D_f = 1.5$ mm) was smaller than the diameter of the openings in the nozzle diaphragms ($D_a = 2$ mm). Comparing the two series of measurements shown in Figs.8.25 and 8.26 indicates that although a very similar dependence on the expansion ratio of the ele-

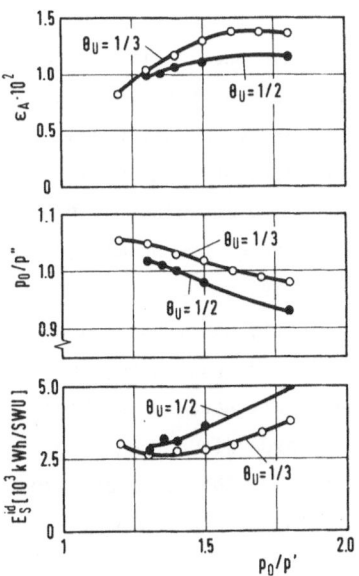

Fig.8.26. Elementary effect ε_A of isotope separation, expansion ratio p_0/p'' of the heavy fraction for setting the UF_6 cut to $\theta_u = 1/3$ and to $\theta_u = 1/2$, and specific energy consumption E_s^{id} versus expansion ratio p_0/p' of the light fraction. Results of separation experiments using an H_2/UF_6 mixture with low UF_6 molar fraction in an annular separation nozzle operated at high UF_6 cuts θ_u; $N_u^0 = 0.02$, $p_0 = 5.5$ mbar, $D_a = 2$ mm, $D_f = 1.5$ mm, $B_a = 1.2$ mm, $B_f = 0.75$ mm

mentary effect of isotope separation is seen, there is a fundamentally different stagnation behavior. In the measurements shown in Fig.8.25, the expansion ratio p_0/p'' of the heavy fraction, which is required to set the UF_6 cut to $\theta_u = 1/4$, rises with p_0/p'. By way of contrast, the expansion ratio of the heavy fraction required to set $\theta_u = 1/2$ and $1/3$ drops with increasing expansion ratio of the light fraction under the conditions pertaining to the measurements shown in Fig.8.26. In this case, the pressure p'' in the heavy fraction is raised clearly beyond the inlet pressure p_0, demonstrating the low pressure losses in the annular separation nozzle. In a cascade consisting of annular separation nozzles, compression of the heavy fraction would not at all be necessary under these conditions.

The low values of ε_A at high expansion ratios are not only the consequence of more intensive mixing in the stagnation zone of the radially converging flow. It is also important to note that UF_6 is more and more concentrated close to the axis of the system with increasing expansion ratio and decreasing UF_6 molar fraction. As a consequence, the stagnation pressure on the axis rises with increasing expansion ratio and, at the same time, the UF_6 molar stream surfaces located concentric around the axis will contract more and more strongly. Similarly, the average dynamic pressure of a gas stream enclosed by a stream surface concentric to the axis will increase with the expansion ratio and with the UF_6 cut appertaining to this stream surface. However, an effective conversion of the dynamic pressure of such a gas stream into static pressure by stagnation and, hence, a strong increase in the pressure of the heavy fraction p'' is feasible only if the diameter D_f of the skimmer opening is not large compared with the diameter D_s of the stream surface enclosing this gas stream. If this condition is not met, i.e., if $D_s \ll D_f$, p'' may be higher than the stagnation pressure of the flow regions further away from the axis and a backflow may occur at the circumference of the skimmer opening. Such backflow from the heavy into the light fraction has been observed in free molecular probe measurements on an annular separation nozzle; it results in a decrease of ε_A especially at high expansion ratios, because in this case the dynamic pressure drops quickly with increasing distance from the axis of rotation. Likewise, it is plausible that high stagnation pressures can be attained especially if the diameter of the skimmer opening is small and the UF_6 cut is high. In addition, a relatively low UF_6 molar fraction is advantageous, since it results in a high UF_6 speed ratio and thus in a pronounced concentration of UF_6 close to the axis of the annular separation nozzle.

Since only a small number of systematic series of experiments have so far been conducted, it is not yet possible to consider the technical expenditure required for uranium isotope separation by the annular separation nozzle. This is even more true of the multi-annular separation nozzle shown in Fig.8.13, with which only a few tentative experiments have been carried out to date.

9. Conclusions

The studies of the physics of the separation nozzle method summarized in this monograph indicate that the flow and diffusion processes in the separation nozzle can be described rather well by the Navier-Stokes equations and by the standard kinetic theory for ternary gaseous mixtures based on the Chapman-Enskog approximation of the Boltzmann equation. The model concepts developed on this basis and greatly improved by extensive experiments make it possible to assess, in a physically plausible way, the influences of the different operating parameters on isotope separation and technical expenditure. Moreover, the importance of these model concepts lies in the fact that new approaches can be derived from them towards the further advancement of the separation nozzle process. This has been demonstrated by improvements achieved to date, such as the biradial nozzle geometry and the double-deflection system.

A quantitative theoretical description of the diffusion and flow processes in the separation nozzle is not possible with the means now available, because major changes of state occur within a few mean free paths in broad areas of the separation nozzle flow. The resultant non-equilibrium effects, such as different temperatures and velocities of the components of the mixture and multimodal velocity distributions, cannot be taken into account by the standard diffusion equations and the Navier-Stokes equations. Therefore, such phenomena have so far been studied mainly experimentally, especially by means of the free molecular probe technique. The Monte-Carlo calculations performed in recent times have not yet furnished any results beyond the findings derived from the experiments /41-44/. According to the present state of the art it may be assumed that the non-equilibrium effects have a predominantly adverse effect on isotope separation. Velocity slip results in a less effective acceleration of UF_6 by the auxiliary gas and a greater increase in entropy in the gaseous mixture, and also the mutual penetration of gas streams in dynamic flow deflection includes a remixing process. However, an effect positive for separation would be a more effective deceleration of the light isotope in a stagnation zone.

Such non-equilibrium effects as well as the measuring techniques developed for analyzing those effects in separation nozzles are of interest not only to aerodynamic separation methods. Similar phenomena occur in many areas of fluid dynamics, molecular physics and gas kinetics; typical examples are the rapid expansion of gaseous mixtures in gasdynamic lasers, the generation of high-energy molecular beams by means of the seeded-beam technique, the behavior of gaseous mixtures in shock waves, and high-speed flows with evaporation of wall material. In each case, large changes of state occur within a few mean free paths and, analogous to the separation nozzle, various molecular species are involved which may differ considerably with respect to molecular weights. Obviously, different velocities and different temperatures of the components of the mixture can occur and the local molecular velocity distribution can be strongly perturbated. Thus, fundamental studies on the flow and diffusion processes in separation nozzles may be relevant to other areas in physics and, vice versa, studies on the topics mentioned above may contribute to an improved understanding of the phenomena typical of the separation nozzle method.

Nevertheless, the main goals of physics development work on the separation nozzle method consist in the elaboration of approaches for further improvement of the separating characteristics and verification of such approaches by separation experiments. Therefore, experiments and phenomenological assessments will continue to be much more important than fundamental theoretical studies on the flow and diffusion processes in separation nozzles. Since the success of this development work can ultimately be gauged only by technical utilization of the results achieved, the physics problems will continue to be influenced very much by the technological implications associated with industrial application of the separation nozzle process.

A factor of particular interest for the further development of the separation nozzle method is the reduction in specific energy consumption because, at the present development status of the method, a major percentage of the separative work cost in commercial separation nozzle facilities will be due to power consumption and expenditure for the electrical driving equipment and the cooling systems. The significance of this aspect is underlined by the fact that recent technological developments in the fabrication of technical separation nozzle systems have allowed the characteristic dimensions of the nozzles to be reduced considerably and the operating pressure to be raised correspondingly. In addition, a clear reduction in the specific fabrication costs of the separation nozzle elements is to be expected /82/. Such progress in fabrication technology reduces the specific investment costs associated with the specific suction volume and the specific slit length and, hence, their contributions to the separative work costs. However, the costs due to the specific energy consumption are not affected by improvements in the fabrication of technical separation nozzles but can be reduced only by improvements in the separation characteristics.

The prospects of further reductions in specific energy consumption and, at the same time, also in the other specific process parameters appear quite positive if one looks at the physical mechanisms characteristic of the separation nozzle method. Unlike the gaseous diffusion method, in which the maximum enrichment factor is equal to the ratio of the mean thermal velocities of the isotopic molecules and, accordingly, to the square root of the ratio of molecular weights /3/, there is no equivalent upper limit in the separation nozzle method for isotope separation. At a given UF_6 cut, the elementary effect of isotope separation can be increased by the same measure as the relative spatial displacement between the isotopic mixture and the auxiliary gas can be enhanced. In particular, the change in state of the mixture associated with the separation process is only partly irreversible in the separation nozzle method while in the gaseous diffusion method there is completely irreversible throttling in the molecular effusion through the microporous separating membrane. The expenditure required to produce the separating centrifugal field and the compression work to be performed on the gaseous mixture can be further reduced in the separation nozzle method, if the dissipative losses in the flow are reduced or the kinetic energy of the gas flows extracted from the separation nozzle is used for pressure recovery or, as in the double-deflection system, for further isotope separation.

Since the reversible thermodynamic work of separation is many orders of magnitude smaller than the corresponding compression work performed on the process gas of the separation nozzle /2/, and since the separation nozzle principle encompasses a large variety of configurations for generating curved flows, no simple lower limit of the specific energy consumption of the separation nozzle process can be indicated. Therefore, it can be assumed that in the long term the separation nozzle process will be competitive both with the gaseous diffusion process and the centrifuge process because of its comparatively simple technology and the resultant relatively low specific capital costs.

References

1   E.W. Becker: "Isotopentrennung", in *Kerntechnik*, ed. by W. Riezler, W. Walcher
    (Teubner, Stuttgart 1958) pp. 227-261;
    H. London, ed.: *Separation of Isotopes* (George Newnes, London 1961);
    H. Barwich: "Isotopentrennung", in *Lehrbuch der Kernphysik*, ed. by. G. Hertz
    (Edition Leipzig/Dausien, Hanau 1963) pp. 55-87
2   W. Ehrfeld, U. Ehrfeld: "Anreicherung von ^{235}U", in *Gmelin Handbuch der Anor-
    ganischen Chemie, URAN*, Ergänzungsband A 2, 57-149 (Springer, Berlin, Heidel-
    berg, New York 1980)
3   D. Massignon: "Gaseous Diffusion", in *Uranium Enrichment*, ed. by S. Villani,
    Topics in Applied Physics, Vol. *35* (Springer, Berlin, Heidelberg, New York 1979)
    pp. 55-182
4   W. Groth: "Gaszentrifugenanlagen zur Anreicherung von Uran-235", Naturwissen-
    schaften *60*, 57-64 (1973)
5   E.W. Becker: "Separation Nozzle", in *Uranium Enrichment*, ed. by S. Villani,
    Topics in Applied Physics, Vol. *35* (Springer, Berlin, Heidelberg, New York
    1979) pp. 245-268
6   A.J.A. Roux, W.L. Grant, R.A. Barbour, R.S. Loubser, J.J. Wannenburg: "Develop-
    ment and Progress of the South African Enrichment Project", in *Nuclear Power
    and its Fuel Cycle*, Vol. *3* (IAEA, Vienna 1977) pp. 171-182
7   D. Gourisse, J.C. Guais: "L'enrichissement de l'uranium per échange chimique",
    Entropie *78*, 42-47 (1977)
8   C.P. Robinson, R.J. Jensen: "Laser Methods of Uranium Isotope Separation, in
    Uranium Enrichment, ed. by S. Villani, Topics in Applied Physics, Vol. *35*,
    (Springer, Berlin, Heidelberg, New York 1979) pp. 269-290
9   E.W. Becker, W. Bier, W. Ehrfeld, K. Schubert, R. Schütte, D. Seidel: "Physics
    and Technology of Separation Nozzle Process", in *Nuclear Energy Maturity*, Proc.
    Eur. Nucl. Conf., Paris 1975, Vol. *12* (Pergamon, Oxford 1976) pp. 44-52
10 E.W. Becker, W. Bier, W. Ehrfeld, D. Seidel, K. Schubert, R. Schütte: "Uranium
    Enrichment by the Separation Nozzle Process", Naturwissenschaften *63*, 407-411
    (1976)
11 E.W. Becker: "The Separation Nozzle Process for Uranium Isotope Enrichment",
    in *Nuclear Power and its Fuel Cycle*, Vol. 3 (IAEA, Vienna 1977) pp. 161-170
12 E.W. Becker: "The Separation Nozzle Process for Enrichment of Uranium-235",
    Prog. Nucl. Energy *1*, 27-39 (1977)
13 P.A.I. Tahourdin: "Final Report on the Jet Separation Method", Rpt. AERE-Br-694,
    Oxford (1946)
14 E.W. Becker, R. Schütte: "Entmischung der Uranisotope mit der Trenndüse",
    Z. Naturforsch. A *11*, 679-680 (1956)
15 E.W. Becker, R. Schütte: "Das Trenndüsenverfahren. III. Entmischung der Uran-
    isotope", Z. Naturforsch. A *15*, 336-347 (1960)
16 E.W. Becker, K. Bier, W. Bier: "Trenndüsenverfahren mit leichtem Zusatzgas",
    Z. Naturforsch. A *17*, 778-785 (1962)
17 E.W. Becker, K. Bier, W. Bier, R. Schütte: "Trenndüsenentmischung der Uraniso-
    tope bei Verwendung leichter Zusatzgase", Z. Naturforsch. A *18*, 246-250 (1963)

18 E.W. Becker, K. Bier, W. Bier, R. Schütte, D. Seidel: "Separation of the Iso-
    topes of Uranium by the Separation Nozzle Process", Angew. Chem. Int. Edit.
    Engl. *6*, 507-518 (1967)
19 E.W. Becker, W. Bier, R. Schütte: "Principles and Economic Aspects of the Sepa-
    ration Nozzle Process", KFK 853, Karlsruhe (1968)
20 E.W. Becker, W. Bier, G. Frey, R. Schütte: "Trenndüsen-Demonstrationsanlage für
    die Urananreicherung", Atomwirtsch. Atomtech. *14*, 249-251 (1969)
21 E.W. Becker:"Die Verfahren zur Trennung der Uranisotope", Kerntechnik *11*,
    129-139 (1969)
22 L.R. Powers, D.J. Roberts, E. von Halle: "An Economic Evaluation of the Sepa-
    ration Nozzle Process for the Production of Enriched Uranium", Rpt. K-OA-2017,
    Oak Ridge (1971)
23 E.W. Becker, W. Bier, W. Ehrfeld, G. Eisenbeiß, G. Frey, H. Geppert, P. Happe,
    G. Heeschen, R. Lücke, D. Plesch, K. Schubert, R. Schütte, D. Seidel, U. Sieber,
    H. Völcker, F. Weis: "The Separation Nozzle Process for Enrichment of U-235",
    in *Peaceful Uses of Atomic Energy*, Vol. *9* (A/CONF 49/P-383, IAEA, Vienna 1972)
    pp. 3-13
24 E.W. Becker, C. Syllus Martins Pinto, H. Völcker: "Development and Technical
    Implementation of the Separation Nozzle Process for Enrichment of Uranium 235",
    Trans. Am. Nucl. Soc. *27*, 477 (1977) and KFK 2678, Karlsruhe (1977)
25 E.W. Becker, P. Nogueira Batista, H. Völcker: "Uranium Enrichment by the Sepa-
    ration Nozzle Method within the Framework of German/Brazilian Cooperation",
    Nucl. Technol. *52*, 105-114 (1981)
26 H. Geppert, P. Schuhmann, U. Sieber, H.P. Stermann, H. Völcker, G. Weinhold:
    "The Industrial Implementation of the Separation Nozzle Process", Int. Conf.
    Uranium Isotope Separation, paper 3, London (1975)
27 E.W. Becker, W. Bier, W. Fritz, P. Happe. D. Plesch, K. Schubert, R. Schütte,
    D. Seidel: "Current Status of Separation Nozzle Technology", Int. Conf. Uranium
    Isotope Separation, paper 2, London (1975)
28 E.W. Becker, W. Bier, K. Schubert, R. Schütte, D. Seidel, U. Sieber:"Technologi-
    cal Aspects of the Separation Nozzle Process", AIChE 69th Annu. Meeting, Chicago
    1976, AIChE Symp. Ser. *73*, 25-29 (1977)
29 E.W. Becker, W. Bier, P. Bley, U. Ehrfeld, W. Ehrfeld, G. Eisenbeiß: "Das Ent-
    wicklungspotential des Trenndüsenverfahrens zur Uran-235-Anreicherung", Atom-
    wirtsch. Atomtech. *18*, 524-527 (1973) and Reaktortagung 1973 des Deutschen
    Atomforums und der Kerntechnischen Gesellschaft, Tagungsband der Zentralstelle
    für Atomkernenergie-Dokumentation (ZAED), 255-258 (1973)
30 E.W. Becker, W. Bier, W. Ehrfeld, K. Schubert, R. Schütte, D. Seidel: "Present
    State and Development Potential of Separation Nozzle Process", KFK 2067, Karls-
    ruhe (1974)
31 E.W. Becker, W. Berkhahn, P. Bley, U. Ehrfeld, W. Ehrfeld, U. Knapp: "Physics
    and Development Potential of the Separation Nozzle Process", Int. Conf. Uranium
    Isotope Separation, paper 1, London (1975)
32 W. Ehrfeld: "The Separation Nozzle Process"; Lecture Series on Aerodynamic Sep-
    aration of Gases and Isotopes, (Von Karman Institute for Fluid Dynamics, Rhode-
    Saint-Genêse 1976)
33 E.W. Becker, P. Bley, U. Ehrfeld, W. Ehrfeld: "Rarefied Gas Dynamics of the
    Separation Nozzle - An Aerodynamic Device for Large-Scale Enrichment of Uranium",
    in *Proceedings of the 10th International Symposium on Rarefied Gas Dynamics*,
    Vol. *1*, ed. by J.L. Potter (American Institute of Aeronautics and Astronautics,
    New York 1977) pp. 3-16
34 W. Ehrfeld: "The Separation Nozzle Process"; Lecture Series on Aerodynamic Sep-
    aration of Gases and Isotopes - Recent Advances (Von Karman Institute for Fluid
    Dynamics, Rhode-Saint-Genêse 1978)
35 E.W. Becker, W. Bier, W. Ehrfeld, G. Eisenbeiß: "Die physikalischen Grundlagen
    der Uran-235-Anreicherung nach dem Trenndüsenverfahren. - Die Wirkung des leich-
    ten Zusatzgases"; Z. Naturforsch. A *26*, 1377-1384 (1971)

36 P. Grassmann: *Physikalische Grundlagen der Verfahrenstechnik* (Sauerländer, Aarau 1970)

37 K. Cohen: *The Theory of Isotope Separation as Applied to the Large-Scale Production of U-235* (McGraw-Hill, New York 1951)

38 M. Benedict, T.H. Pigford: *Nuclear Chemical Engineering* (McGraw-Hill, New York 1957)

39 S. Villani: "Isotope Separation"; American Nuclear Society, Hinsdale, Ill. (1976)

40 E. von Halle: "Stage Separation Theory"; Lecture Series on Aerodynamic Separation of Gases and Isotopes - Recent Advances (Von Karman Institute for Fluid Dynamics, Rhode-Saint-Genèse 1978)

41 A.U. Chatwani, M. Fiebig, N.K. Mitra, W. Schwan, P. Bley, W. Ehrfeld, W. Fritz: "Tracer Monte Carlo Simulation for an Isotope Separation Nozzle", in *Proceedings of the 12th International Symposium on Rarefied Gas Dynamics*, ed. by S.S. Fisher (American Institute of Aeronautics and Astronautics, New York 1981)

42 R.L. Fox, R.R. Eaton: "Investigation of Nozzles, Jets, and Channels for the Separation of Heavy Isotopes", in *Proceedings of the 10th International Symposium on Rarefied Gas Dynamics*, Vol. *1*, ed. by J.L. Potter (American Institute of Aeronautics and Astronautics, New York 1977) pp. 35-48, cf. also SAND 76-0004, SAND 76-0598 and SAND 75-0176, Albuquerque (1976/1977)

43 A.U. Chatwani, M. Fiebig, N.K. Mitra, W. Ehrfeld: "Nonequilibrium Effects and Their Modeling in Separation Nozzles", in *Proceedings of the 12th International Symposium on Rarefied Gas Dynamics*, ed. by S.S. Fisher (American Institute of Aeronautics and Astronautics, New York 1981)

44 R.R. Eaton, R.L. Fox, K.J. Touryan: "Isotope Enrichment by Aerodynamic Means: A Review and Some Theoretical Considerations"; J. Energy *1*, 229-236 (1977)

45 U. Ehrfeld, W. Ehrfeld: "Untersuchungen zum Einfluß von Wärmetransportvorgängen auf den Strömungsverlauf und die Entmischung in der Trenndüse mit molekular angeströmten Druck- und Temperatursonden"; KFK 1634, Karlsruhe (1972)

46 P. Kirch, R. Schütte: "Messungen des Thermodiffusionsfaktors und Bestimmung des molekularen Wechselwirkungspotentials von gasförmigem Uranhexafluorid"; Z. Naturforsch. A *22*, 1532-1537 (1967)

47 S.I. Sandler, E.A. Mason: "Kinetic-Theory Deviations from Blanc's Law of Ion Mobilities"; J. Chem. Phys. *48*, 2873-2875 (1968)

48 E.P. Ney, F.C. Armistead: "The Self-Diffusion Coefficient of Uranium Hexafluoride"; Phys. Rev. *71*, 14-19 (1947)

49 S. Ljungren: "The Diffusion of Uranium Hexafluoride in Some Other Gases"; Ark. Kemi *24*, 1-46 (1965)

50 K. Bier, H. Brandstädter, U. Ehrfeld, W. Ehrfeld: "Untersuchungen des Strömungsverlaufs von He/Ar- und He/SF$_6$-Gemischen in der Trenndüse mit molekular angeströmten Drucksonden"; KFK 1440, Karlsruhe (1971)

51 W. Berkhahn, W. Ehrfeld, G. Krieg: "Berechnung der Uranisotopenentmischung in der Trenndüse bei kleinen UF$_6$-Molenbrüchen im Zusatzgas"; KFK 2351, Karlsruhe (1976)

52 W. Berkhahn, W. Ehrfeld, G. Krieg: "Influence of Flowfield Structure on Uranium Isotope Separation in the Separation Nozzle"; Nucl. Technol. *40*, 329-340 (1978)

53 G.F. Malling, E. von Halle: "Aerodynamic Isotope Separation Processes for Uranium Enrichment: Process Requirements"; Rpt. K-OA-2872, Oak Ridge (1976)

54 T.W. Kao: "Baro-Diffusion Effect on Diffusion Time"; Phys. Fluids *10*, 1814-1817 (1967)

55 U. Ehrfeld, W. Ehrfeld, E. Schmid: "Free Molecular Probe Measurements of Isotope Distribution in the Separation Nozzle", in *Proceedings of the 11th International Symposium on Rarefied Gas Dynamics*, Vol. *1*, ed. by R. Campargue (Commissariat à l'Energie Atomique, Paris 1979) pp. 617-627

56 R. Dürr: "Theoretische Untersuchung einer viskosen, kompressiblen Strömung in einer gekrümmten Düse"; KFK 1630, Karlsruhe (1972)

57 P. Bley, R. Dürr, W. Ehrfeld, G. Eisenbeiß: "Die physikalischen Grundlagen der Uran-235-Anreicherung nach dem Trenndüsenverfahren. III. Die Mechanismen der

Entropieerzeugung in der Trenndüsenströmung"; Z. Naturforsch. A *28*, 1273-1280 (1973)

58 K. Bier, W. Ehrfeld: "Zur Untersuchung von Strömungen verdünnter Gase mit Miniatur-Drucksonden"; Z. Angew. Phys. *28*, 70-77 (1969) and KFK 842, Karlsruhe (1969)

59 K. Bier, H. Brandstädter, W. Ehrfeld: "Miniatur-Drucksonden zur Untersuchung von Strömungs- und Diffusionsvorgängen in verdünnten Gasgemischen"; Z. Angew. Phys. *29*, 205-208 (1970)

60 G.N. Patterson: "Theory of Free-Molecule, Orifice Type Pressure Probes in Isentropic and Nonisentropic Flows"; UTIA-Report 41, Toronto (1959)

61 P.C. Hughes: "Theory for the Free-Molecule Impact Probe at an Arbitrary Angle of Attack"; UTIAS-Report 103, Toronto (1965)

62 H. Breton, W. Ehrfeld, G. Krieg: "Selective Measurements of Density and Rotational Temperature in Gaseous Mixtures Using a Tunable CO_2-Laser", in *Proceedings of the 11th International Symposium on Rarefied Gas Dynamics*, Vol. *1*, ed. by R. Campargue (Commissariat à l'Energie Atomique, Paris 1979) pp. 545-552

63 H. Breton, W. Ehrfeld, G. Krieg: "Laser Diagnostics of Opposed He/C_7F_{14} Jets", in *Proceedings of the 12th International Symposium on Rarefied Gas Dynamics*, ed. by S.S. Fisher (American Institute of Aeronautics and Astronautics, New York 1981) pp. 590-597

64 P. Bley, W. Ehrfeld: "Bestimmung der dissipativen Verluste in der Trenndüsenströmung mit Pitot-Sonden"; KFK 1562, Karlsruhe (1972)

65 H.M. Mott-Smith: "The Solution of the Boltzmann Equation for a Shock Wave"; Phys. Rev. *82*, 885-892 (1951)

66 P. Bley, U. Ehrfeld, W. Ehrfeld: "Enhancement of Nozzle Discharge Coefficients in Rarefied Flows of Disparate Mass Mixtures", in *Proceedings of the 11th International Symposium on Rarefied Gas Dynamics*, Vol. *1*, ed. by R. Campargue (Commissariat à l'Energie Atomique, Paris 1979) pp. 241-252

67 F.R. Block: "Die Multifluidtheorie in der Thermodynamik irreversibler Prozesse"; Thesis, TH Aachen (1970)

68 W. Ehrfeld, E. Schmid: "Untersuchungen mit molekular angeströmten Drucksonden zum Strömungsverlauf in einem Trenndüsensystem mit Nachtrennung der schweren Fraktion durch eine zusätzliche mechanische Strahlumlenkung"; KFK 2004, Karlsruhe (1974)

69 E.W. Becker, W. Bier, P. Bley, U. Ehrfeld, W. Ehrfeld, G. Eisenbeiß, F.J. Rosenbaum, E. Schmid: "Die physikalischen Grundlagen der Uran-235-Anreicherung nach dem Trenndüsenverfahren. IV. Trenndüsensystem mit zweifacher Strahlumlenkung und trifraktionärer Gasabsaugung"; Z. Naturforsch. A *32*, 401-410 (1977)

70 P. Bley, W. Ehrfeld, U. Heiden: "Uranisotopenentmischung in der Trenndüse bei Rückstau der schweren Fraktion"; KFK 2580, Karlsruhe (1978)

71 U. Ehrfeld, W. Ehrfeld, E. Schmid: "Untersuchungen mit molekular angeströmten Sonden zum räumlichen Verlauf der Isotopenentmischung in der Trenndüse bei Rückstau der schweren Fraktion"; KFK 2724, Karlsruhe (1978)

72 W. Berkhahn, P. Bley, H. Breton, W. Ehrfeld: "Einfluß der Verfahrensgastemperatur auf die Entmischung der Uranisotope in der Trenndüse"; KFK 2466, Karlsruhe (1977)

73 W. Bier, G. Eisenbeiß, G. Heeschen: "Die physikalischen Grundlagen der Uran-235-Anreicherung nach dem Trenndüsenverfahren. II. Vergleich der leichten Zusatzgase H_2, He und D_2"; Z. Naturforsch. A *28*, 1267-1272 (1973)

74 P. Bley, W. Ehrfeld, F.M. Jäger, U. Knapp: "Entwicklung und Erprobung einer Versuchsapparatur für die Optimierung von Trenndüsensystemen zur Anreicherung von Uran-235"; KFK 2092, Karlsruhe (1975)

75 P. Bley, W. Ehrfeld, F.M. Jäger, U. Knapp: "Meßtechnik und Versuchsstrategie bei der Optimierung von Trenndüsensystemen zur Anreicherung von Uran-235"; Reaktortagung 1975 des Deutschen Atomforums und der Kerntechnischen Gesellschaft, Tagungsband der Zentralstelle für Atomkernenergie-Dokumentation (ZAED), 322-325 (1975)

76 P. Bley, W. Ehrfeld, U. Knapp, G. Krieg: "Entwicklung von Gasanalysatoren für das Trenndüsenverfahren in Zusammenarbeit mit der Industrie"; KFK-Nachrichten *11*,

Heft 2, 48-57 (1979)

77 W. Ehrfeld, U. Knapp: "Experimentelle Untersuchung der Uranisotopentrennung in einem Trenndüsensystem mit gegenseitiger Umlenkung von zwei frontal aufeinander gerichteten Gasstrahlen"; KFK 2138, Karlsruhe (1975)

78 U. Ehrfeld: "Separation Nozzle Systems with Dynamic Flow Deflection"; Lecture Series on Aerodynamic Separation of Gases and Isotopes - Recent Advances (Von Karman Institute for Fluid Dynamics, Rhode-Saint-Genêse 1978)

79 E.W. Becker, P. Bley, U. Ehrfeld, W. Ehrfeld, U. Knapp: "Enrichment of ^{235}U by Separation Nozzle Systems with Dynamic Flow Deflection"; Trans. Am. Nucl. Soc. *28*, 371-372 (1978)

80 E.W. Becker, P. Bley, U. Ehrfeld, W. Ehrfeld, U. Knapp, G. Krieg: "Separation Nozzle Systems with Dynamic Flow Deflection", in *Proceedings of the 11th International Symposium on Rarefied Gas Dynamics*, Vol. *1*, ed. by R. Campargue (Commissariat à l'Energie Atomique, Paris 1979) pp. 587-600

81 P. Bley, W. Ehrfeld: "Molecular Dynamics of Disparate Mass Mixtures in Opposed Jets", in *Proceedings of the 12th International Symposium on Rarefied Gas Dynamics*, ed. by S.S. Fisher (American Institute of Aeronautics and Astronautics, New York 1981) pp. 577-589

82 E.W. Becker, W. Bier, W. Ehrfeld, K. Schubert, D. Seidel: "Entwicklung und technische Anwendung des Trenndüsenverfahrens zur Anreicherung von Uran-235"; KFK-Nachrichten *13*, Heft 1-2, 50-57 (1981)

Subject Index

S.A. Losev

Gasdynamic Laser

1981. 100 figures. X, 297 pages. (Springer
Series in Chemical Physics, Volume 12)
ISBN 3-540-10503-4

The gas dynamic Laser represents a higher-
power source of coherent and directional
radiation due to a rapidly cooling gas. The
principles and recent improvements are described. The basic aspects of vibrational relaxation kinetics and relevant physicochemical
processes are reviewed, and the fundamental
dynamics of relaxing gases in a nozzle are
given. Mathematical models for the operation
mechanism of gas dynamic lasers are presented and problems relating to the creation of
new recombinations and to plasma-dynamic
lasers are discussed.

Hydrodynamic Instabilities and the Transition to Turbulence

Editors: H. L. Swinney, J. P. Gollub
With contributions by numerous experts
1981. 81 figures. XII, 292 pages. (Topics in
Applied Physics, Volume 45)
ISBN 3-540-10390-2

Major advances in hydrodynamic stability
and the transition to turbulence are reviewed
by prominent physicists, engineers, and
mathematicians in this volume. It contains
chapters on bifurcation theory, strance attrac-
tors, Rayleigh-Bénard convection, circular
Couette flow, shear flow instabilities, geo-
physical fluid dynamics, and other chaotic
systems. This book, written at a level accessible to graduate students and nonspecialists,
provides the first overview of the subject
incorporating the substantial progress of the
past decade. Since nonlinear dynamical phenomena in many areas of science are closely
related, this volume will be of interest to
mathematicians, condensed-matter physicists,
engineers, chemists, and others concerned
with stability and chaotic behavior.

Modelling of Chemical Reaction Systems

Proceedings of an International Workshop,
Heidelberg, Federal Republic of Germany,
September 1–5, 1980

Editors: K. H. Ebert, P. Deuflhard, W. Jäger
1981. 163 figures. X, 389 pages. (Springer
Series in Chemical Physics, Volume 18)
ISBN 3-540-10983-8

The purpose of the workshop from which the
proceedings resulted war to bring together
engineers, mathematicians and chemists on
the problems of chemical reactions.
Numerical-mathematical and analytical-
mathematical methods for chemical reaction
systems are discussed, and are given for these
methods in science and engineering. New
methods for the description of the dynamical
behaviour of chemical reaction systems are
given. The problems of large and very stiff
differential equations, which result from chemical systems, are treated. The proceedings
give a survey of the most modern methods
for treating chemical reaction systems, the
problems which are now being studied by
scientists all over the world, and the future
aspects of this field of research.

Springer-Verlag
Berlin
Heidelberg
New York

Nonlinear Phenomena in Chemical Dynamics

Proceedings of an International Conference, Bordeaux, France, September 7–11, 1981

Editors: **C. Vidal, A. Pacault**
1981. 124 figures. X, 280 pages. (Springer Series in Synergetics, Volume 12)
ISBN 3-540-11294-4

Contents: General Nonlinear Behavior. – Weak Turbulence. – Stochastic Analysis. – Critical Phenomena. – Coupling of Oscillators. – Reaction-Diffusion Problems. – Biochemical Processes. – From Bistability to Oscillations. – Mathematical Modeling. – Poster Abstracts. – Index of Contributors.

Turbulence

Editor: **P. Bradshaw**
With contributions by numerous experts
2nd corrected and updated edition. 1978.
47 figures, 4 tables. XI, 339 pages. (Topics in Applied Physics, Volume 12)
ISBN 3-540-08864-4

There are several books which survey turbulence in depth, but none which adequately treats it in depth as the most important fluid-dynamic phenomenon in engineering and the earth sciences. This book is a unified treatment of most of the turbulence problems of aeronautical, mechanical, and chemical engineering, meteorology and oceanography. Each chapter is written by an expert in one of these disciplines, but emphasizes phenomena rather than hardware details so as to make the material accessible to non-specialists. As well as descriptions of phenomena, the book contains detailed discussions of methods for calculating turbulent flow fields and heat transfer.

Turbulent Reacting Flows

Editors: **P. A. Libby, F. A. Williams**
1980. 38 figures, 3 tables. XI, 243 pages
(Topics in Applied Physics, Volume 44)
ISBN 3-540-10192-6

This volume provides the background material necessary for unraveling the complexities and interactions arising in turbulent combustion. It also sumarizes recent theoretical approaches that have been developed in the field.
In a book devoted to such a specialized field such as turbulent reacting flows appropriate background knowledge must be assumed. Thus some knowledge of fluid mechanical turbulence and of aerothermochemistry is presumed. With this knowledge as a foundation, the volume is self-contained and may be used as a textbook for a graduate course in mechanics, applied physics or engineering or as an introduction to this new research area. Throughout, the exposition emphasizes the fundamental bases and theoretical aspects of each topic, providing a perspective of, stimulating interest in, a challenging and important field.

Uranium Enrichment

Editor: **S. Villani**
1979. 140 figures, 25 tables. XI, 322 pages
(Topics in Applied Physics, Volume 35)
ISBN 3-540-09385-0

Contents: *S. Villani:* Review of Separation Processes. – *B. Brigoli:* Cascade Theory. – *D. Massignon:* Gaseous Diffusion. – *Soubbaramayer:* Centrifugation. – *E. W. Becker:* Separation Nozzle. – *C. P. Robinson, R. J. Jensen:* Laser Methods of Uranium Isotope Separation. – *F. Boeschoten, N. Nathrath:* Plasma Separating Effects. – Additional References with Titles. – Subject Index.

Springer-Verlag Berlin Heidelberg New York